Catalog of
Ford Truck
ID Numbers
1946-1972
Pickup and Ranchero
Compiled by the Staff of Cars & Parts Magazine

Published by
Amos Press Inc.
911 Vandemark Road
Sidney, Ohio 45365

Publishers of
Cars & Parts
The Magazine Serving the Car Hobbyist

Cars & Parts Collectible Series
Muscle Cars of the '60s/'70s
Collectible Trucks
Collector Car Annual
Cars of the '50s
Cars of the '60s

Catalog of American Car ID Numbers 1960-69
Catalog of American Car ID Numbers 1970-79
Catalog of Chevy Truck ID Numbers 1946-72

Salvage Yard Treasures
A Guide to America's Salvage Yards

D1476542

Copyright 1992 by Amos Press Inc.

Distribution by Motorbooks International Publishers and Wholesalers
P.O. Box 2, Osceola WI 54020 USA

Printed and bounded in the United States of America

Library of Congress Cataloging-In-Publication Date
ISBN 1-880524-03-1

ACKNOWLEDGMENTS

The staff of *Cars & Parts* Magazine devoted more than a year to the research and development of the Catalog of Ford Truck ID Numbers 1946-72. It has been a labor-intensive project which required assistance from hundreds of truck collectors, clubs, researchers, and a tremendous amount of research at car and truck shows, swap meets and auctions.

This book wouldn't have been possible without very special help from the following:

Motor Vehicle Manufacturers Association

Dan Kirchner - Researcher

Jim Wirth - Researcher

Automotive History Collection of the Detroit Public Library

Thank you for making this book possible.

Catalog of
Ford Truck
ID Numbers
1946-1972
Pickup and Ranchero
Compiled by the Staff of Cars & Parts Magazine

INTRODUCTION

Authentication has become such a critical issue within the old car/truck hobby that the need for a comprehensive, accurate and dependable identification guide has become quite apparent to anyone involved in buying, selling, restoring, judging, owning, researching or appraising a collector vehicle. With this indepth and detailed ID guide, the staff of Cars & Parts magazine has compiled as much data as possible on the years covered to help take the fear out of buying a collector truck.

Deciphering paint codes, verifying vehicle identification numbers (VIN), interpreting body codes and authenticating engine numbers will become a much easier process with this guide at your side. Putting this previously obscure information at your fingertips has not been a simple task, but one worth the tremendous investment, time and money spent on its production.

The information contained in the Catalog of Ford Truck ID Numbers 1946-72 was compiled from a variety of sources including original manufacturers' catalogs (when available) and official shop manuals, truck data books, service bulletins, etc. The Cars & Parts staff and researchers made every attempt to verify the information continued herein. However, Ford made changes from year-to-year and model-to-model, as well as during mid-year production. In the immediate years after WW II, the numbering system was a continuation of that used during the war years, making identification difficult during the early post war years. In some instances, conflicting information surfaced during the course of our indepth research. Consequently, Cars & Parts can not guarantee the absolute accuracy of all data presented in this ID catalog.

HOW TO USE THIS CATALOG

SAMPLE: 1968-72 F-SERIES / BRONCO TRUCK TAG

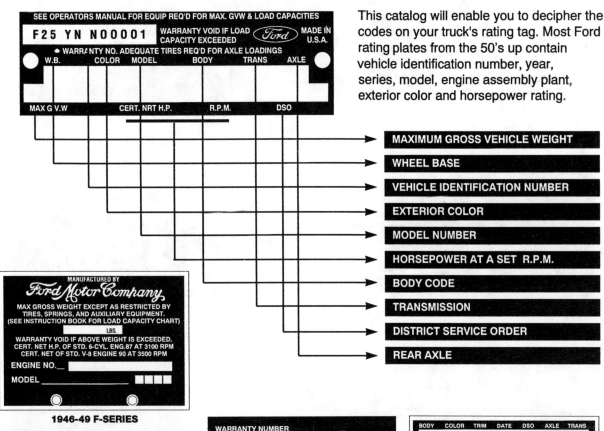

This catalog will enable you to decipher the codes on your truck's rating tag. Most Ford rating plates from the 50's up contain vehicle identification number, year, series, model, engine assembly plant, exterior color and horsepower rating.

- MAXIMUM GROSS VEHICLE WEIGHT
- WHEEL BASE
- VEHICLE IDENTIFICATION NUMBER
- EXTERIOR COLOR
- MODEL NUMBER
- HORSEPOWER AT A SET R.P.M.
- BODY CODE
- TRANSMISSION
- DISTRICT SERVICE ORDER
- REAR AXLE

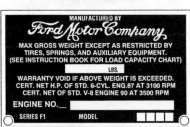

1946-49 F-SERIES

1963-65 F-SERIES

1963-65 COURIER / RANCHERO

1950 F-SERIES

1966-67 F-SERIES / BRONCO

1966-67 COURIER / RANCHERO

1951-52 F-SERIES

1952-59 COURIER / RANCHERO

1968-69 COURIER / RANCHERO

1953-59 F-SERIES

1960-62 F-SERIES

1960-62 COURIER / RANCHERO

1970-72 COURIER / RANCHERO

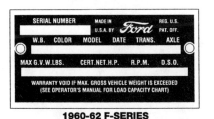

1946 SEDAN DELIVERY

1946 6½ FT. PICKUP

1946 7½ FT. PANEL

1946 8 FT. EXPRESS

1946 LIGHT DUTY PICKUP

SERIAL AND ENGINE NUMBER

    • I G Y   2 2 7 5 2 4 •

The serial number and the engine number are the same on a 1946 Ford truck and are a continuation (with minor modifications) of the sequence in effect prior to World War II. It is stamped into the clutch housing and visible when the transmission cover is removed.

THE FIRST DIGIT in the serial number prefix for all 6-cylinder powered vehicles begins with a 1 (one). The prefix for all 8-cylinder powered trucks begins with 69 (except the sedan delivery which carries the passenger car prefix of 9).

BODY STYLE	ENGINE	PREFIX
Sedan delivery	6	1GA
1/2-Ton	6	1GC
1-Ton	6	1GY
Sedan delivery	8	99A
1/2-Ton	8	699C
1-Ton	8	699Y

THE SECOND AND THIRD DIGITS (6-cylinder only) identify the model.

VEHICLE	MODEL CODE
Sedan delivery	GA
1/2-Ton	GC
1-Ton	GY

THE THIRD AND FOURTH DIGITS (8-cylinder only) identify the model.

VEHICLE	MODEL CODE
Sedan delivery	9A
1/2-Ton pickup	9C
1-Ton	9Y

NOTE: The letter G denotes a 6-cylinder and a 9 indicates an 8-cylinder. Letter designations identify the model line, ie. A - passenger car, C - 1/2-ton pickup and Y - 1-ton.

THE LAST SIX DIGITS indicate the production sequence.

RATING PLATE

The rating plate attached to the glove box door on each model indicates the maximum gross vehicle weight, certified net horsepower at specified r.p.m., engine series and model identification number.

MANUFACTURED BY
Ford Motor Company
MAX GROSS WEIGHT EXCEPT AS RESTRICTED BY TIRES, SPRINGS, AND AUXILIARY EQUIPMENT. (SEE INSTRUCTION BOOK FOR LOAD CAPACITY CHART)
LBS.
WARRANTY VOID IF ABOVE WEIGHT IS EXCEEDED. CERT. NET H.P. OF STD. 6-CYL. ENG.81 AT 3100 RPM CERT. NET OF STD. V-8 ENGINE 87 AT 3500 RPM
ENGINE NO.__   I G Y   2 2 7 5 2 4
MODEL _____

THE MAXIMUM GROSS VEHICLE WEIGHT indicates the maximum gross vehicle weight in pounds.

LIGHT DUTY

SERIES	MODEL NUMBER	NO. CYL.	WHEEL BASE	G.V.W.
Sedan delivery	6GA-78	6	114"	—
Sedan delivery	69C-78	8	114"	—
Pickup	6GC-83	6	114"	4,700
Pickup	69A-83	8	114"	4,700
Panel	6GC-82	6	114"	4,700
Panel	69A-82	8	114"	4,700
Stake	6GC-86	6	114"	4,700
Stake	69A-86	8	114"	4,700
Chassis-cab	6GC-81	6	114"	4,700
Chassis-cab	69A-81	8	114"	4,700
Chassis-cowl	6GC-84	6	114"	4,700
Chassis-cowl	69A-84	8	114"	4,700
Chassis-windshield	6GC-85	6	114"	4,700
Chassis-windshield	69A-85	8	114"	4,700

ONE-TON

Express pickup	6GY-83	6	122"	6,600
Express pickup	69Y-83	8	122"	6,600
Panel	6GY-82	6	122"	6,600
Panel	69Y-82	8	122"	6,600
Stake	6GY-86	6	122"	6,600
Stake	69Y-86	8	122"	6,600
Chassis-cab	6GY-81	6	122"	6,600
Chassis-cab	69Y-81	8	122"	6,600
Chassis-cowl	6GY-84	6	122"	6,600
Chassis-cowl	69Y-84	8	122"	6,600
Chassis-windshield	6GY-85	6	122"	6,600
Chassis-windshield	69Y-85	8	122"	6,600

BODY TYPE

BODY STYLE	CODE
Sedan delivery	78
Closed cab	81
Panel delivery (1-ton)	82
Pickup	83
Chassis/cowl	84
Chassis/windshield	85
Stake	86

REAR AXLE

REAR AXLE	RATIO 6-CYL.	RATIO 8-CYL.
Sedan delivery (std.)	3.78:1	3.54:1
Sedan delivery (opt.)	4.11:1	3.78:1
Light duty (std.)	3.78:1	3.54:1
Light duty (opt.)	3.54:1	3.78:1
Light duty (opt.)	4.11:1	4.11:1
1-Ton (std.)	4.86:1	4.86:1
1-Ton (opt.)	4.11:1	4.11:1

THE PAINT COLOR CODE did not appear on the truck until later years. To aid restorers, the following Ford paint order numbers are provided.

SEDAN DELIVERY

COLOR	CODE
Light Moonstone Gray	M-3981
Navy Blue No. 1*	M-3982
Navy Blue No. 2	M-3982
Botsford Blue Green	M-3983
Modern Blue	M-3987
Dynamic Maroon	M-3989
Greenfield Green	M-3990
Dark Slate Gray	M-3991
Silver Sand Poly	M-3992
Willow Green	M-14140
Black	—

* Used in production until 11-1-45

TRUCKS

COLOR	CODE
Village Green	M-3949
Greenfield Green*	M-3990

* Replaced Village Green 12-1-45

ENGINE SPECIFICATIONS

ENGINE CODE	NO. CYL.	CID	HORSE-POWER	COMP. RATIO	CARB
G	6	226	90	6.7:1	1 BC
V-8	8	239	100	6.75:1	2 BC

* On a V-8 engine with sleeves the engine is marked with the letters "HS" stamped on top of the block beside the inner front corner of the left cylinder head.

1947 9 FT. PANEL

1947 SEDAN DELIVERY

1947 7½ FT. PANEL

1947 8 FT. EXPRESS

1947 6½ FT. PICKUP

SERIAL AND ENGINE NUMBER PLATE

> • 7 1 G Y   2 3 0 6 9 8 •

The serial number and the engine number are the same on a 1947 Ford truck. It is stamped into the clutch housing and visible when the transmission cover is removed.

BODY STYLE	ENGINE	PREFIX
Sedan delivery	6	71GA
1/2-Ton pickup	6	71GC
1-Ton	6	71GY
Sedan delivery*	6	77HA
Sedan delivery	8	799A
1/2-Ton pickup	8	799C
1-Ton	8	799Y

* After September 1947

THE FIRST DIGIT in the serial number is a 7 (seven) for the year 1947.

THE SECOND DIGIT (6-cylinder only) is a 1(one) and is a continuation of the numbering sequence used prior to World War II.

THE THIRD AND FOURTH DIGITS (6-cylinder only) identify the model.

VEHICLE	MODEL CODE
Sedan delivery	GA
1/2-Ton pickup	GC
1-Ton	GY

THE SECOND DIGIT (8-cylinder only) is a 9(nine).

THE THIRD AND FOURTH DIGITS (8-cylinder only) identify the model.

VEHICLE	MODEL CODE
Sedan delivery	9A
1/2-Ton pickup	9C
1-Ton	9Y

THE LAST SIX DIGITS identify the production sequence numbers.

NOTE: The letter G denotes a 6-cylinder and a 9 indicates an 8-cylinder. Letter designations identify the model line, ie. A - passenger car, C - 1/2-ton pickup and Y - 1-ton.

RATING PLATE

The rating plate attached to the glove box door on each model indicates the maximum gross vehicle weight, certified net horsepower at specified r.p.m., engine series and model identification number.

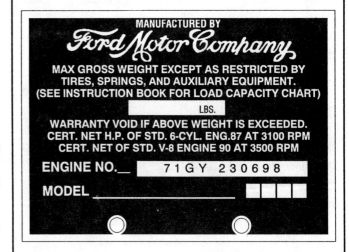

MANUFACTURED BY
Ford Motor Company
MAX GROSS WEIGHT EXCEPT AS RESTRICTED BY TIRES, SPRINGS, AND AUXILIARY EQUIPMENT. (SEE INSTRUCTION BOOK FOR LOAD CAPACITY CHART)
LBS.
WARRANTY VOID IF ABOVE WEIGHT IS EXCEEDED. CERT. NET H.P. OF STD. 6-CYL. ENG.87 AT 3100 RPM CERT. NET OF STD. V-8 ENGINE 90 AT 3500 RPM
ENGINE NO. _   7 1 G Y   2 3 0 6 9 8
MODEL _

THE MAXIMUM GROSS VEHICLE WEIGHT indicates the maximum gross vehicle weight in pounds.

LIGHT DUTY

SERIES	MODEL NO.	NO. CYL.	WHEEL BASE	....G.V.W.
Sedan delivery	7GA-78	6	114"	—
Sedan delivery	79A-78	8	114"	—
Pickup	7GC-83	6	114"	4,700
Pickup	79C-83	8	114"	4,700
Panel	7GC-82	6	114"	4,700
Panel	79C-82	8	114"	4,700
Stake	7GC-86	6	114"	4,700
Stake	79C-86	8	114"	4,700
Platform	7GC-80	6	114"	4,700
Platform	79C-80	8	114"	4,700
Chassis-cab	7GC-81	6	114"	4,700
Chassis-cab	79C-81	8	114"	4,700
Chassis-cowl	7GC-84	6	114"	4,700
Chassis-cowl	79C-84	8	114"	4,700
Chassis-windshield	7GC-85	6	114"	4,700
Chassis-windshield	79C-85	8	114"	4,700

ONE-TON

SERIES	MODEL NO.	NO. CYL.	WHEEL BASE	....G.V.W.
Express pickup	7GY-83	6	122"	6,600
Express pickup	79Y-83	8	122"	6,600
Panel	7GY-82	6	122"	6,600
Panel	79Y-82	8	122"	6,600
Stake	7GY-86	6	122"	6,600
Stake	79Y-86	8	122"	6,600
Platform	7GY-80	6	122"	6,600
Platform	79Y-80	8	122"	6,600
Chassis-cab	7GY-81	6	122"	6,600
Chassis-cab	79Y-81	8	122"	6,600
Chassis-cowl	7GY-84	6	122"	6,600
Chassis-cowl	79Y-84	8	122"	6,600
Chassis-windshield	7GY-85	6	122"	6,600
Chassis-windshield	79Y-85	8	122"	6,600

THE BODY CODE indicates the body type.

BODY STYLE	CODE
Sedan delivery	78
Platform	80
Closed cab	81
Panel delivery	82
Pickup	83
Chassis-cowl	84
Chassis-windshield	85
Stake	86

REAR AXLE

REAR AXLE	RATIO 6-CYL.	RATIO 8-CYL.
Sedan delivery (std.)	3.78:1	3.54:1
Sedan delivery (opt.)	4.11:1	3.78:1
Light duty (std.)	3.78:1	3.54:1
Light duty (opt.)	3.54:1	3.78:1
Light duty (opt.)	4.11:1	4.11:1
1-Ton (std.)	4.86:1	4.86:1
1-Ton (opt.)	4.11:1	4.11:1

THE PAINT COLOR CODE did not appear on the truck until later years. To aid restorers, the following Ford paint order numbers are provided.

SEDAN DELIVERY

COLOR	CODE
Rotunda Gray	M-14220
Barcelona Blue	M-14221
Moonstone Maroon	M-14222
Glade Green	M-14223
Feather Gray	M-14224
Blue Gray Poly	M-14225
Tucson Tan	M-14227
Midland Maroon Poly	M-14202
Shoal Green Gray Poly	M-14228
Strato Blue	M-14201

TRUCKS

COLOR	CODE
Greenfield Green	M-3990
Medium Luster Black	M-1722
Light Moonstone Gray	M-3981
Modern Blue	M-3987
Vermilion	M-1722

ENGINE SPECIFICATIONS

ENGINE CODE	NO. CYL.	CID	HORSE-POWER	COMP. RATIO	CARB
G	6	226	90	6.70:1	1 BC
H*	8	226	95	6.80:1	1 BC
V-8	8	239	100	6.75:1	2 BC

* "H" started September 1947

1948 F-100 PICKUP

1948 SEDAN DELIVERY

1948 9 FT. PANEL

1948 8 FT. EXPRESS

1948 7½ FT. PANEL

1948 6½ FT. PICKUP

SERIAL AND ENGINE NUMBER PLATE

• 88RC 139260 •

The serial number and engine number are the same on the 1948 Ford truck and are located on the glove box door and on the left side of the frame near the steering gear mounting bracket. The first four digits indicate the year, series and number of cylinders.

SERIES/BODY STYLE		ENGINE PREFIX
Sedan delivery	6	88HA
F-1, 1/2-ton	6	87HC
F-2, 3/4-ton	6	87HD
F-3, 3/4-ton HD	6	87HY
F-4, 1-ton	6	87HT
Sedan Delivery	8	889A
F-1, 1/2-ton	8	88RC
F-2, 3/4-ton	8	88RD
F-3, 3/4-ton HD	8	88RY*
F-4, 1 ton	8	88RTL

THE FIRST DIGIT in the serial numer prefix is an 8(eight) for the year 1948.

THE SECOND AND THIRD DIGITS are either 7H, which denotes a 6-cylinder engine or 8H which denotes an 8-cylinder engine — except the sedan delivery which carries a 9(nine), the passenger car style serial number sequence.

THE FOURTH DIGIT is a letter which identifies the model line, ie. A — passenger, C — 1/2 ton and Y — 3/4 ton and the 1 ton which carries a T.

THE LAST SIX DIGITS indicates the production sequence.

* Effective with engine 87H number 110141, Ford replaced the Y with the letter D to identify the F-2 series 3/4-ton truck.

RATING PLATE

The rating plate attached to the glove box door on each model indicates the maximum gross vehicle weight, certified net horsepower at specified r.p.m., engine series and model identification number.

THE MAXIMUM GROSS VEHICLE WEIGHT indicates the maximum gross vehicle weight in pounds.

SERIES	TYPE	G.V.W. (LBS.)
—	Sedan Delivery	—
F1	Conventional	4,700
F2	Conventional	5,700
F3	Conventional	6,800
F4	Conventional (single wheel)	7,500
F4	Conventional (dual wheel)	10,000

SEDAN DELIVERY

TYPE	MODEL NO.	NO. CYL.
Sedan Delivery	87HA	6
Sedan Delivery	87HA	8

F-1 SERIES

TYPE	MODEL NO.	NO. CYL.
Pickup	8HC-83	6
Pickup	8RC-83	8
Panel	8HC-82	6
Panel	8RC-82	8
Stake	8HC-86	6
Stake	8RC-86	8
Platform	8HC-80	6
Platform	8RC-80	8
Chassis-cab	8HC-81	6
Chassis-cab	8RC-81	8

F-1 SERIES (continued)

Chassis-cowl	8HC-84	6
Chassis-cowl	8RC-84	8
Chassis-windshield	8HC-85	6
Chassis-windshield	8RC-85	8

F-2 SERIES

TYPE	MODEL NO.	NO. CYL.
Express pickup	8HD-83	6
Express pickup	8RD-83	8
Stake	8HD-86	6
Stake	8RD-86	8
Platform	8HD-80	6
Platform	8RD-80	8
Chassis-cab	8HD-81	6
Chassis-cab	8RD-81	8
Chassis-cowl	8HD-84	6
Chassis-cowl	8RD-84	8
Chassis-windshield	8HD-85	6
Chassis-windshield	8RD-85	8

F-3 SERIES

TYPE	MODEL NO.	NO. CYL.
Express pickup	8HY-83	6
Express pickup	8RY-83	8
Stake	8HY-86	6
Stake	8RY-86	8
Platform	8HY-80	6
Platform	8RY-80	8
Chassis-cab	8HY-81	6
Chassis-cab	8RY-81	8
Chassis-cowl	8HY-84	6
Chassis-cowl	8RY-84	8
Chassis-windshield	8HY-85	6
Chassis-windshield	8RY-85	8

F-4 SERIES

TYPE	MODEL NO.	NO. CYL.
Stake	8HTL-86	6
Stake	8RTL-86	8
Platform	8HTL-80	6
Platform	8RTL-80	8
Chassis-cab	8HTL-81	6
Chassis-cab	8RTL-81	8
Chassis-cowl	8HTL-84	6
Chassis-cowl	8RTL-81	8
Chassis-windshield	8HTL-85	6
Chassis-windshield	8RTL-85	8

THE BODY CODE indicates the body type.

BODY STYLE	CODE
Sedan delivery	78
Platform	80
Closed cab	81
Panel delivery	82
Pickup	83
Chassis-cowl	84
Chassis-windshield	85
Stake	86

REAR AXLE

SERIES	STANDARD	OPTIONAL
F-1	3.73:1	4.27:1
F-2, F-3	4.86:1	4.11:1
F-4	5.14:1	5.83:1
F-4	—	6.67:1

THE PAINT COLOR CODE did not appear on the truck until later years. To aid restorers, the following Ford paint numbers are provided.

TRUCK

COLOR	CODE
Vermilion	M-1722
Meadow Green	M-14283
Chrome Yellow	M-14301
Birch Gray	M-14286
Black	M-1724

SEDAN DELIVERY

COLOR	CODE
Rotunda Gray	M-14220
Barcelona Blue	M-14221
Moonstone Maroon	M-14222
Parrot Green Poly	M-14226
Taffy Tan	M-14242
Glade Green	M-14223
Feather Gray	M-14224
Blue Gray Poly	M-14225
Tuscon Tan	M-14227
Maize Yellow	M-14229
Pheasant Red	M-14230
Midland Maroon Poly	M-14202
Shoal Green Grey Poly	M-14228
Strato Blue	M-14201

ENGINE SPECIFICATIONS

ENGINE CODE	NO. CYL.	CID	HORSE-POWER	COMP. RATIO	CARB
7HT	6	226	95	6.8:1	1 BC
8R	8	239	100	6.8:1	2 BC

SERIAL AND ENGINE NUMBER PLATE

• 98RC 139260 •

The serial number and engine number are the same on the 1949 Ford truck and are located on the glove box door and on the left side of the frame near the steering gear mounting bracket. The first 4 digits indicate the year, series and number of cylinders.

SERIES / BODY STYLE	ENGINE	PREFIX
F1, 1/2-ton	6	97HC
F2, 3/4-ton	6	97HY
F3, 3/4-ton	6	97HY
F4, 1 ton	6	97HT
F1, 1/2-ton	8	98RC
F2, 3/4-ton	8	98RY
F3, 3/4-ton	8	98RY
F4, 1 ton	8	98RT

THE FIRST DIDGIT in the serial number is a 9(nine) for the year 1949.

THE SECOND AND THIRD DIGITS are either 7H, which indicates a 6-cylinder engine or 8R, which denotes an 8-cylinder engine.

THE FOURTH DIGIT is a letter designation which identifies the model line, ie. A — passenger, C — 1/2 ton, Y — 3/4 ton and T — 1ton truck.

THE LAST SIX DIGITS indicate the production sequence.

RATING PLATE

The rating plate attached to the glove box door on each model indicates the maximum gross vehicle weight, certified net horsepower at specified r.p.m., engine series and model identification number.

THE MAX. G.V.W. (LBS.) CODE indicates the maximum gross vehicle weight in pounds.

SERIES	TYPE	GVW (LBS.)
F1	Conventional	4,700
F2	Conventional	5,700
F3	Conventional	6,800
F4	Conventional (single wheel)	7,500
F4	Conventional (dual wheel)	10,000

F-1 SERIES

SERIES	MODEL NO.	NO. CYL.
Pickup	9HC-83	6
Pickup	9RC-83	8
Panel	9HC-82	6
Panel	9RC-82	8
Stake	9HC-86	6
Stake	9RC-86	8
Platform	9HC-80	6
Platform	9RC-80	8
Chassis-cab	9HC-81	6
Chassis-cab	9RC-81	8
Chassis-cowl	9HC-84	6
Chassis-cowl	9RC-84	8
Chassis-windshield	9HC-85	6
Chassis-windshield	9HC-85	8

F-2 SERIES

SERIES	MODEL NO.	NO. CYL.
Express pickup	9HD-83	6
Express pickup	9RD-83	8
Stake	9HD-86	6
Stake	9RD-86	8
Platform	9HD-80	6
Platform	9RD-80	8
Chassis-cab	9HD-81	6
Chassis-cab	9RD-81	8
Chassis-cowl	9HD-84	6
Chassis-cowl	9RD-84	8
Chassis-windshield	9HD-85	6
Chassis-windshield	9RD-85	8

F-3 SERIES

SERIES	MODEL NO.	NO. CYL.
Express pickup	9HY-83	6
Express pickup	9RY-83	8
Stake	9HY-86	6
Stake	9RY-86	8
Platform	9HY-80	6
Platform	9RY-80	8
Chassis-cab	9HY-81	6
Chassis-cab	9RY-81	8
Chassis-cowl	9HY-84	6
Chassis-cowl	9RY-84	8
Chassis-windshield	9HY-85	6
Chassis-windshield	9RY-85	8

F-4 SERIES

SERIES	MODEL NO.	NO. CYL.
Stake	9HTL-86	6
Stake	9RTL-86	8
Platform	9HTL-80	6
Platform	9RTL-80	8
Chassis-cab	9HTL-81	6
Chassis-cab	9RTL-81	8
Chassis-cowl	9HTL-84	6
Chassis-cowl	9RTL-84	8
Chassis-windshield	9HTL-85	6
Chassis-windshield	9RTL-85	8

THE BODY CODE indicates the body type.

BODY STYLE	CODE
Platform	80
Closed cab	81
Panel delivery	82
Pickup	83
Chassis-cowl	84
Chassis-windshield	85
Stake	86

REAR AXLE

SERIES	STANDARD	OPTIONAL
F-1	3.73:1	4.27:1
F-2, F-3	4.86:1	4.11:1
F-4	5.14:1	5.83:1
F-4	—	6.67:1

THE PAINT COLOR CODE did not appear on the truck until later years. To aid restorers, the following Ford paint order numbers are provided.

COLOR	CODE
Med. Luster Black	M-1724
Vermilion	M-1722
Meadow Green	M-14283
Birch Gray	M-14286
Chrome Yellow	M-14301

ENGINE SPECIFICATIONS

ENGINE CODE	NO. CYL.	CID	HORSE-POWER	COMP. RATIO	CARB
7HT	6	226	95	6.8:1	1 BC
8R	8	239	100	6.8:1	2 BC

1950 F-1 PICKUP

1950 F-1 PICKUP

SERIAL AND ENGINE NUMBER PLATE

● 98RC 139260 ●

The serial number and engine number found on the 1950 Ford truck are a continuation of the 1949 numbers. A plate containing a series of numbers and letters is located on the glove box door. The number is also found on the left side of the frame near the steering gear mounting bracket. See 1949 for details.

SERIES / BODY STYLE	ENGINE	PREFIX
F1, 1/2-ton	6	97HC
F2, 3/4-ton	6	97HY
F3, 3/4-ton	6	97HY
F4, 1 ton	6	97HT
F1, 1/2-ton	8	98RC
F2, 3/4-ton	8	98RY
F3, 3/4-ton	8	98RY
F4, 1 ton	8	98RT

RATING PLATE

The rating plate attached to the dispatch box door on each model indicates the maximum gross vehicle weight, certified net horsepower at specified r.p.m., engine series and model identification number.

THE MAXIMUM GROSS VEHICLE WEIGHT (LBS.) CODE indicates the maximum gross vehicle weight in pounds.

SERIES	TYPE	GVW (LBS.)
F1	Conventional	4,700
F2	Conventional	5,700
F3	Conventional	6,800
F4	Conventional (single wheel)	7,500
F4	Conventional (dual wheel)	10,000

F-1 SERIES

SERIES	MODEL NO.	NO. CYL.
Pickup	9HC-83	6
Pickup	9RC-83	8
Panel	9HC-82	6
Panel	9RC-82	8
Stake	9HC-86	6
Stake	9RC-86	8
Platform	9HC-80	6
Platform	9RC-80	8
Chassis-cab	9HC-81	6
Chassis-cab	9RC-81	8
Chassis-cowl	9HC-84	6
Chassis-cowl	9RC-84	8
Chassis-windshield	9HC-85	6
Chassis-windshield	9RC-85	8

F-2 SERIES

SERIES	MODEL NO.	NO. CYL.
Express pickup	9HD-83	6
Express pickup	9RD-83	8
Stake	9HD-86	6
Stake	9RD-86	8
Platform	9HD-80	6
Platform	9RD-80	8
Chassis-cab	9HD-81	6
Chassis-cab	9RD-81	8
Chassis-cowl	9HD-84	6
Chassis-cowl	9RD-84	8
Chassis-windshield	9HD-85	6
Chassis-windshield	9RD-85	8

F-3 SERIES

SERIES	MODEL NO.	NO. CYL.
Express pickup	9HY-83	6
Express pickup	9RY-83	8
Stake	9HY-86	6
Stake	9RH-86	8
Platform	9HY-80	6
Platform	9RY-80	8
Chassis-cab	9HY-81	6
chassis-cab	9RY-81	8
Chassis-cowl	9HY-84	6
Chassis-cowl	9RY-84	8
Chassis-windshield	9HY-85	6
Chassis-windshield	9RY-85	8

F-4 SERIES

SERIES	MODEL NO.	NO. CYL.
Stake	9HTL-86	6
Stake	9RTL-86	8
Platform	9HTL-80	6
Platform	9RTL-80	8
Chassis-cab	9HTL-81	6
Chassis-cab	9RTL-81	8
Chassis-cowl	9HTL-84	6
Chassis-cowl	9RTL-84	8
Chassis-windshield	9HTL-85	6
Chassis-windshield	9RTL-85	8

THE BODY CODE indicates the body type.

BODY STYLE	CODE
Platform	80
Closed cab	81
Panel delivery	82
Pickup	83
Chassis-cowl	84
Chassis-windshield	85
Stake	86

REAR AXLE

SERIES	STANDARD	OPTIONAL
F-1	3.73:1	4.27:1
F-2,F-3	4.86:1	4.11:1
F-4	5.14:1	5.83:1
F-4	—	6.67:1

THE PAINT COLOR CODE did not appear on the truck until later years. To aid restorers, the following numbers are provided.

COLOR	CODE
Vermilion	M-1722
Raven Black	M-1724
Palisade Green	M-14341
Meadow Green	M-14283
Sheridan Blue	M-14285
Silvertone Gray	M-14197
Birch Gray	M-14286
Dover Gray	M-14344
Sunland Beige	M-14343
Prime	M-4415

ENGINE SPECIFICATIONS

ENGINE CODE	NO. CYL.	CID	HORSE-POWER	COMP. RATIO	CARB
7HT	6	226	95	6.8:1	1 BC
8R	8	239	100	6.8:1	2 BC

1951 FORD TRUCK (TILL SEPTEMBER 15, 1951)

SERIAL AND ENGINE NUMBER PLATE

● 98RC 139260 ●

Till September 15, 1951 the serial number and engine number found on the 1951 Ford truck are a continuation of the 1949 numbers. A plate containing a series of numbers and letters is located on the glove box door. The number is also found on the left side of the frame near the steering gear mounting bracket. See 1949 for details.

SERIES / BODY STYLE	ENGINE	PREFIX
F1, 1/2-ton	6	97HC
F2, 3/4-ton	6	97HY
F3, 3/4-ton	6	97HY
F4, 1 ton	6	97HT
F1, 1/2-ton	8	98RC
F2, 3/4-ton	8	98RY
F3, 3/4-ton	8	98RY
F4, 1 ton	8	98RT

RATING PLATE

The rating plate attached to the glove box door on each model indicates the maximum gross vehicle weight, certified net horsepower at specified r.p.m., engine series and model identification number.

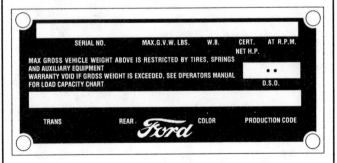

THE MAXIMUM GROSS VEHICLE WEIGHT indicates the maximum gross vehicle weight in pounds.

SERIES	TYPE	G.V.W. (LBS.)
F1	Conventional	4,700
F2	Conventional	5,700
F3	Conventional	6,800
F4	Conventional (single wheel)	7,500
F4	Conventional (dual wheel)	10,000

F-1 SERIES

SERIES	MODEL NO.	NO. CYL.
Pickup	9HC-83	6
Pickup	9RC-83	8
Panel	9HC-82	6
Panel	9RC-82	8
Deluxe panel	9HC-82B	6
Deluxe panel	9RC-82B	8
Stake	9HC-86	6
Stake	9RC-86	8
Platform	9HC-80	6
Platform	9RC-80	8
Chassis-cab	9HC-81	6
Chassis-cab	9RC-81	8
Chassis-cowl	9HC-84	6
Chassis-cowl	9RC-84	8
Chassis-windshield	9HC-85	6
Chassis-windshield	9RC-85	8

F-2 SERIES

SERIES	MODEL NO.	NO. CYL.
Express pickup	9HD-83	6
Express pickup	9RD-83	8
Stake	9HD-86	6
Stake	9RD-86	8
Platform	9HD-80	6
Platform	9RD-80	8
Chassis-cab	9HD-81	6
Chassis-cab	9RD-81	8
Chassis-cowl	9HD-84	6
Chassis-cowl	9RD-84	8
Chassis-windshield	9HD-85	6
Chassis-windshield	9RD-85	8

F-3 SERIES

SERIES	MODEL NO.	NO. CYL.
Express pickup	9HY-83	6
Express pickup	9RY-83	8
Stake	9HY-86	6
Sake	9RY-86	8
Platform	9HY-80	6
Platform	9RY-80	8
Chassis-cab	9HY-81	6
Chassis-cab	9RY-81	8
Chassis-cowl	9HY-84	6
Chassis-cowl	9RY-84	8
Chassis-windshield	9HY-85	6
Chassis-windshield	9RY-85	8

F-4 SERIES

SERIES	MODEL NO.	NO. CYL.
Stake	9HTL-86	6
Stake	9RTL-86	8
Platform	9HTL-80	6
Platform	9RTL-80	8
Chassis-cab	9HTL-81	6
Chassis-cab	9RTL-81	8
Chassis-cowl	9HTL-84	6
Chassis-cowl	9RTL-84	8
Chassis-windshield	9HTL-85	6
Chassis-windshield	9RTL-85	8

THE BODY CODE indicates the body type.

BODY STYLE	CODE
Platform	80
Closed cab	81
Panel delivery deluxe	82A
Panel delivery	82
Pickup	83
Chassis-cowl	84
Chassis-windshield	85
Stake	86

THE REAR AXLE CODE indicates the ratio of the rear axle installed in the vehicle.

SERIES	STANDARD	OPTIONAL
F1	3.92:1	4.27:1
F2, F-3	4.86:1	4.11:1
F-4	5.14:1	5.83:1
F-4	—	6.67:1

ENGINE SPECIFICATIONS

ENGINE CODE	NO. CYL.	CID	HORSE-POWER	COMP. RATIO	CARB
7HT	6	226	95	6.8:1	1 BC
8R	8	239	100	6.8:1	2 BC

F-SERIES (AFTER SEPTEMBER 15, 1951)

RATING PLATE

The information indicated on the rating plate on vehicles built after September 4, 1951, is the vehicle identification number, maximum gross vehicle weight (lbs.), wheelbase, certified net horsepower at r.p.m., transmission, rear axle, exterior color, production code and D.S.O. numbers.

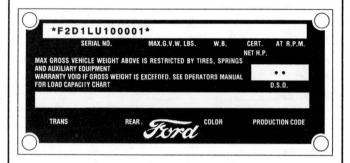

THE VEHICLE IDENTIFICATION NUMBER is a
series of letters and numbers on the rating plate. The VIN number identifies the series, engine, model year, assembly plant and production sequence.

FIRST AND SECOND DIGITS: Identify the series

SERIES	CODE
F-1 conventional	F1
F-2 conventional	F2
F-3 conventional	F3
F-4 conventional	F4

THIRD DIGIT: Identifies the engine

ENGINE	CODE
226 cid, 6 cyl.	H
239 cid, 8 cyl.	R

FOURTH DIGIT: Identifies the model year (1951)

FIFTH AND SIXTH DIGITS: Identify the assembly plant

ASSEMBLY PLANT	CODE
Atlanta, GA	AT
Buffalo, NY	BF
Chester, PA	CS
Chicago, IL	CH
Dallas, TX	DI
Dearborn, MI	DA
Edgewater, NJ	EG
Highland Park, MI	HM
Kansas City, MO	KC
Long Beach, CA	LB
Louisville, KY	LU
Memphis, TN	MP
Norfolk, VA	NR
Richmond, VA	RH
Somerville, MA	SR
Twin City (St. Paul), MN	SP

LAST SIX DIGITS: Identify the consecutive unit number

THE MAX. G.V.W. (LBS.) CODE indicates the maximum gross vehicle weight in pounds.

SERIES	TYPE	G.V.W. (LBS.)
F1	Conventional	4,700
F2	Conventional	5,700
F3	Conventional	6,800
F4	Conventional (single wheel)	7,500
F4	Conventional (dual wheel)	10,000

THE CERT. NET H.P. AT R.P.M. CODE indicates the certified net horsepower at the specified r.p.m.

CYLINDER	HORSEPOWER
6	87 HP @ 3100 r.p.m.
8	90 HP @ 3500 r.p.m.

THE D.S.O. CODE indicates the domestic special order.

THE TRANSMISSION CODE indicates the transmission type installed in the vehicle.

TYPE	CODE
3-Speed regular	3
3-Speed heavy duty	3HD
4-Speed spur	4
4-Speed synchronized	4SYN

THE REAR AXLE CODE indicates the ratio of the rear axle installed in the vehicle.

MODEL	STANDARD	OPTIONAL	OVERDRIVE
F-1	3.92:1	4.27:1	4.09:1
F-2, F-3	4.86:1	4.11:1	
F-4	5.14:1	5.83:1	

THE EXTERIOR COLOR CODE indicates the paint color used on the vehicle.

COLOR	CODE
Black	A
Sheridan Blue	B
Alpine Blue	D
Sea Island Green	G
Silvertone Gray	H
Meadow Green	M
Vermilion	N
Prime	P
Special paint	SS

THE PRODUCTION CODE indicates the day of the month, the month and the truck produced.

MONTH	FIRST YEAR	SECOND YEAR
January	A	N
February	B	P
March	C	Q
April	D	R
May	E	S
June	F	T
July	G	U
August	H	V
September	J	W
October	K	X
November	L	Y
December	M	Z

ENGINE SPECIFICATIONS

ENGINE CODE	NO. CYL.	CID	HORSE-POWER	COMP. RATIO	CARB
H	6	226	95	6.8:1	1 BC
R	8	239	100	6.8:1	2 BC

F-SERIES
RATING PLATE

The information indicated on the rating plate is the vehicle identification number, maximum gross vehicle weight (lbs.), the wheelbase, certified net horsepower at r.p.m., D.S.O., transmission type, rear axle, exterior color and production code.

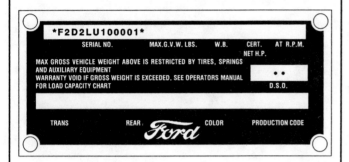

THE VEHICLE IDENTIFICATION NUMBER is a series of letters and numbers on the rating plate. The VIN number identifies the series, engine, model year, assembly plant and production sequence.

FIRST AND SECOND DIGITS: Identify the series

SERIES	CODE
F-1 conventional	F1
F-2 conventional	F2
F-3 conventional	F3
F-4 conventional	F4

THIRD DIGIT: Identifies the engine

ENGINE	CODE
215 cid, 6 cyl.	D
239 cid, 8 cyl.	R

FOURTH DIGIT: Identifies the model year (1952)

FIFTH AND SIXTH DIGITS: Identify the assembly plant

ASSEMBLY PLANT	CODE
Atlanta, GA	AT
Buffalo, NY	BF
Chester, PA	CS
Chicago, IL	CH
Dallas, TX	DL
Dearborn, MI	DA
Edgewater, NJ	EG
Highland Park, MI	HM
Kansas City, MO	KC
Long Beach, CA	LB
Louisville, KY	LU
Memphis, TN	MP
Norfolk, VA	NR
Richmond, VA	RH
Somerville, MA	SR
Twin City (St. Paul), MN	SP

LAST SIX DIGITS: Identify the consecutive unit number

THE VEHICLE DATA appears on the two lines following the vehicle identification number.

THE MAX. G.V.W. (LBS.) CODE indicates the maximum gross vehicle weight in pounds.

SERIES	TYPE	G.V.W. (LBS.)
F1	Conventional	4,700
F2	Conventional	5,700
F3	Conventional	6,800
F4	Conventional single wheel	7,500
F4	Conventional dual wheel	10,000

THE W.B. (WHEELBASE) CODE indicates the wheelbase in inches.

THE CERT. NET H.P. AT R.P.M. CODE indicates the certified net horsepower at the specified r.p.m.

CYLINDER	R.P.M.
6	91 @ 3400 r.p.m.
8	96 @ 3400 r.p.m.

THE D.S.O. CODE indicates the domestic special order.

THE TRANSMISSION CODE indicates the transmission type installed in the vehicle.

TYPE	CODE
3-Speed regular	3
3-Speed heavy duty	3HD
4-Speed spur	4
4-Speed synchronized	4SYN

THE REAR AXLE CODE indicates the ratio of the rear axle installed in the vehicle.

MODEL	STANDARD	OPTIONAL	OVERDRIVE
F-1	3.92:1	4.27:1	4.09:1
F-2, F-3	4.86:1	4.11:1	
F-4	5.14:1	5.83:1	

THE EXTERIOR COLOR CODE indicates the paint color used on the vehicle.

COLOR	CODE
Black	A
Sheridan Blue	B
Woodsmoke Gray	D
Glenmist Green	G
Sandpiper Tan	K
Meadow Green	N
Vermilion	M
Prime	P
Special paint	SS

THE PRODUCTION CODE indicates the day of the month, the month and the truck produced.

MONTH	FIRST YEAR	SECOND YEAR
January	A	N
February	B	P
March	C	Q
April	D	R
May	E	S
June	F	T
July	G	U
August	H	V
September	J	W
October	K	X
November	L	Y
December	M	Z

ENGINE SPECIFICATIONS

ENGINE CODE	NO. CYL.	CID	HORSE-POWER	COMP. RATIO	CARB
D	6	215	101	7.0:1	1 BC
R	8	239	106	6.8:1	2 BC

COURIER
RATING PLATE

The information indicated on the rating plate is the vehicle identification number, body type, exterior color, interior trim and production code.

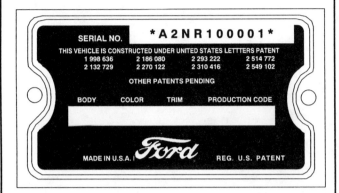

THE VEHICLE IDENTIFICATION NUMBER is a series of letters and numbers on the rating plate. The VIN number identifies the engine, model year, assembly plant and production sequence.

FIRST DIGIT: Identifies the engine

ENGINE	CODE
215 cid, 6 cyl.	A
239 cid, 8 cyl.	B

SECOND DIGIT: Identifies the model year (1952)

THIRD AND FOURTH DIGITS: Identify the assembly plant

ASSEMBLY PLANT	CODE
Atlanta, GA	AT
Buffalo, NY	BF
Chester, PA	CS
Chicago, IL	CH
Dallas, TX	DL
Dearborn, MI	DA
Edgewater, NJ	EG
Kansas City, MO	KC
Long Beach, CA	LB
Louisville, KY	LU
Memphis, TN	MP
Norfolk, VA	NR
Richmond, VA	RH
Somerville, MA	SR
Twin City (St. Paul), MN	SP

LAST SIX DIGITS: Identify the consecutive unit number

THE VEHICLE DATA appears on the line following the vehicle identification number.

THE BODY CODE indicates the body type.

BODY TYPE	CODE
Custom delivery	78A

THE EXTERIOR COLOR CODE indicates the paint color used on the vehicle.

COLOR	CODE
Raven Black	A
Sheridan Blue	B
Alpine Blue	C
Woodsmoke Gray	D
Shannon Green Metallic	E
Meadowbrook Green	F
Glenmist Green	G
Carnival Red	H
Hawaiian Bronze	J
Sandpiper Tan	K
Prime	P
Special paints	SS

THE INTERIOR TRIM CODE indicates the key to the trim color and material used on the vehicle.

COLOR	VINYL	CLOTH	LEATHER	CODE
Maroon/Gray		•		A
Tan		•		B,F
Tan	•			S
Gray/Red		•		D
Green		•		E
Black/Red	•		•	G
Brown/Tan	•		•	H
Brown/Tan	•			N
Green/Ivory	•		•	J
Dk. Blue/Lt. Blue	•		•	K
Green/Gray	○	•		M
Blue/Gray	•	•		R
Mahogany/Straw	•			T
Dk. Blue/Ivory			•	U
Dk. Brown/Ivory			•	V
Dk. Brown/Gray	•			X

THE PRODUCTION CODE indicates the day of the month, the month and the truck produced.

MONTH	FIRST YEAR	SECOND YEAR
January	A	N
February	B	P
March	C	Q
April	D	R
May	E	S
June	F	T
July	G	U
August	H	V
September	J	W
October	K	X
November	L	Y
December	M	Z

ENGINE SPECIFICATIONS

ENGINE CODE	NO. CYL.	CID	HORSE-POWER	COMP. RATIO	CARB
A	6	215	101	7.0:1	1 BC
B	8	239	110	7.2:1	2 BC

1953 F-100 PICKUP

F-SERIES
RATING PLATE

The information indicated on the rating plate is the vehicle identification number, maximum gross vehicle weight (lbs.), wheelbase, certified net horsepower at r.p.m., transmission type, rear axle, exterior color, production code and D.S.O. numbers.

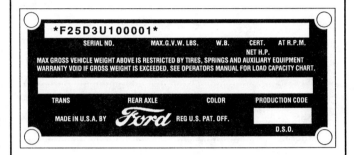

THE VEHICLE IDENTIFICATION NUMBER is a series of letters and numbers on the rating plate. The VIN number identifies the series, engine, model year, assembly plant and production sequence.

FIRST, SECOND AND THIRD DIGITS: Identify the series

SERIES	CODE
F-100	F10
F-250	F25
F-350	F35

FOURTH DIGIT: Identifies the engine

ENGINE	CODE
215 cid, 6 cyl.	D
239 cid, 8 cyl.	R

FIFTH DIGIT: Identifies the model year (1953)

SIXTH DIGIT: Identifies the assembly plant

ASSEMBLY PLANT	CODE
Atlanta, GA	A
Buffalo, NY	B
Chester, PA	C
Chicago, IL	G
Dallas, TX	D
Dearborn, MI	F
Edgewater, NJ	E
Highland Park, MI	H
Kansas City, KS	K
Long Beach, CA	L
Louisville, KY	U
Memphis, TN	M
Norfolk, VA	N
Richmond, VA	R
Somerville, MA	S
Twin City (St. Paul), MN	P

LAST SIX DIGITS: Identify the consecutive unit number

THE VEHICLE DATA appears on the two lines following the vehicle identification number.

THE MAX. G.V.W. LBS. CODE identifies the maximum gross vehicle weight in pounds.

SERIES	TYPE	G.V.W. (LBS.)
F100	Conventional	4,800
F250	Conventional	6,900
F350	Conventional (single wheel)	7,700
F350	Conventional (dual wheel)	9,500

THE W.B. (WHEELBASE) CODE identifies the wheelbase of the vehicle in inches.

THE CERT. NET H.P. AT R.P.M. CODE identifies the certified net horsepower at the specified r.p.m.

THE TRANSMISSION CODE identifies the transmission type installed in the vehicle.

TYPE	CODE
3-Speed regular 3
3-Speed heavy duty...3HD
3-Speed automatic...AUTO
3-Speed overdrive 3OD
4-Speed synchronized...4SYN

THE REAR AXLE CODE identifies the ratio of the rear axle installed in the vehicle.

RATIO

3.92:1 ....

4.09:1 ....

4.27:1 ....

4.86:1 ....

5.14:1 ....

5.83:1 ....

THE EXTERIOR COLOR CODE indicates the paint color used on the vehicle.

COLOR	CODE
Raven Black ....A
Sheridan Blue..B
Glacier Blue...C
Woodsmoke Gray ....D
Seafoam Green..G
Vermillion (Torch Red) ....N
Meadow Green...R
Prime ....P
Special ....SS

THE PRODUCTION CODE indicates the day of the month, the month and the car produced.

MONTH	FIRST YEAR	SECOND YEAR
January ....A.............................N
February..........................B.............................P
March ....C.............................Q
April ....D.............................R
May ....E.............................S
June ....F.............................T
July..................................G.............................U
August ....H.............................V
September........................J.............................W
October ....K.............................X
November.........................L.............................Y
December.........................M.............................Z

THE D.S.O. CODE indicates the domestic special order.

ENGINE SPECIFICATIONS

ENGINE CODE	NO. CYL.	CID	HORSE-POWER	COMP. RATIO	CARB
D	6	215	101	7.0:1	1 BC
R	8	239	110	7.2:1	2 BC

COURIER
RATING PLATE

The information indicated on the rating plate is the vehicle identification number, body type, exterior color, interior trim and production code.

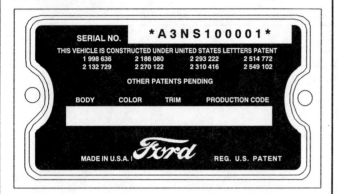

THE VEHICLE IDENTIFICATION NUMBER is a series of letters and numbers on the rating plate. The VIN number identifies the engine, model year, assembly plant, body style and production sequence.

FIRST DIGIT: Identifies the engine

ENGINE	CODE
215 cid, 6 cyl.	A
239 cid, 8 cyl.	B

SECOND DIGIT: Identifies the model year (1953)

THIRD DIGIT: Identifies the assembly plant

ASSEMBLY PLANT	CODE
Atlanta, GA	A
Buffalo, NY	B
Chester, PA	C
Dallas, TX	D
Edgewater, NJ	E
Dearborn, MI	F
Chicago, IL	G
Highland Park, MI	H
Kansas City, KS	K
Long Beach, CA	L
Memphis, TN	M
Norfolk, VA	N
Twin City (St. Paul), MN	P
Richmond, VA	R
Somerville, MA	S
Louisville, KY	U

FOURTH DIGIT: Identifies the body style

BODY STYLE	CODE
Sedan delivery	S

LAST SIX DIGITS: Identify the consecutive unit number

THE VEHICLE DATA appears on the line following the vehicle identification number.

THE BODY CODE indicates the body type.

BODY TYPE	CODE
Mainline	78A

THE EXTERIOR COLOR CODE indicates the paint color used on the vehicle.

COLOR	CODE
Raven Black	A
Sheridan Blue	B
Glacier Blue	C
Woodsmoke Gray	D
Timberline Green Metallic	E
Fernmist Green	F
Seafoam Green	G
Carnival Red Metallic	H
Polynesian Bronze Metallic	J
Sandpiper Tan	K
Sungate Ivory	L
Coral Flame Red	M
Prime	P
Special paints	SS

THE INTERIOR TRIM CODE indicates the key to the trim color and material used on the vehicle.

COLOR	CODE
Dk. Brown vinyl	U
Pigskin vinyl	N

THE PRODUCTION CODE indicates the day of the month, the month and the car produced.

MONTH	FIRST YEAR	SECOND YEAR
January	A	N
February	B	P
March	C	Q
April	D	R
May	E	S
June	F	T
July	G	U
August	H	V
September	J	W
October	K	X
November	L	Y
December	M	Z

ENGINE SPECIFICATIONS

ENGINE CODE	NO. CYL.	CID	HORSE-POWER	COMP. RATIO	CARB
A	6	215	101	7.0:1	1 BC
B	8	239	110	7.2:1	2 BC

F-SERIES
RATING PLATE

The information indicated on the rating plate is the vehicle identification number, maximum gross vehicle weight (lbs.), wheelbase, certified net horsepower at r.p.m., transmission type, rear axle, exterior color, production code, and D.S.O. numbers.

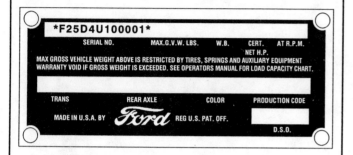

THE VEHICLE IDENTIFICATION NUMBER is a series of letters and numbers on the rating plate. The VIN identifies the series, engine, model year, assembly plant, and production number.

FIRST, SECOND AND THIRD DIGITS: Identify the series

SERIES	CODE
F-100	F10
F-250	F25
F-350	F35

FOURTH DIGIT: Identifies the engine

ENGINE	CODE
223 cid, 6 cyl.	D
239 cid, 8 cyl.	R

FIFTH DIGIT: Identifies the model year (1954)

SIXTH DIGIT: Identifies the assembly plant

ASSEMBLY PLANT	CODE
Atlanta, GA	A
Buffalo, NY	B
Chester, PA	C
Chicago, IL	G
Dallas, TX	D
Dearborn, MI	F
Edgewater, NJ	E
Highland Park, MI	H
Kansas City, KS	K
Long Beach, CA	L
Louisville, KY	U
Memphis, TN	M
Norfolk, VA	N
Richmond, VA	R
Somerville, MA	S
Twin City (St. Paul), MN	P

LAST SIX DIGITS: Identify the consecutive unit number

THE VEHICLE DATA appears on the two lines following the vehicle identification number.

THE MAX. G.V.W. LBS. CODE indicates the maximum gross vehicle weight in pounds.

SERIES	TYPE	G.V.W. (LBS.)
F100	Conventional	4,800
F250	Conventional	6,900
F350	Conventional (single wheel)	7,700
F350	Conventional (dual wheel)	9,500

THE W.B. (WHEELBASE) CODE indicates the wheelbase of the vehicle in inches.

THE CERT. NET H.P. AT R.P.M. CODE indicates the certified net horsepower at the specified r.p.m.

THE TRANSMISSION CODE indicates the transmission type installed in the vehicle.

TYPE	CODE
3-Speed regular	3
3-Speed heavy duty	3HD
3-Speed automatic	AUTO
3-Speed overdrive	3OD
4-Speed synchronized	4SYN

THE REAR AXLE CODE indicates the ratio of the rear axle installed in the vehicle.

RATIO

3.92:1 ....
4.09:1 ....
4.27:1 ....
4.86:1 ....
5.14:1 ....
5.83:1 ....

THE EXTERIOR COLOR CODE indicates the paint color used on the vehicle.

COLOR	CODE
Raven Black	A
Sheridan Blue	B
Glacier Blue	D
Sea Haze Green	H
Vermillion (Torch Red)	R
Meadow Green	U
Goldenrod Yellow	V
Prime	P
Special	SS

THE PRODUCTION CODE indicates the day of the month, the month and the car produced.

MONTH	FIRST YEAR	SECOND YEAR
January	A	N
February	B	P
March	C	Q
April	D	R
May	E	S
June	F	T
July	G	U
August	H	V
September	J	W
October	K	X
November	L	Y
December	M	Z

THE D.S.O. CODE indicates the domestic special order.

ENGINE SPECIFICATIONS

ENGINE CODE	NO. CYL.	CID	HORSE-POWER	COMP. RATIO	CARB
D	6	223	115	7.2:1	1 BC
R	8	239	130	7.2:1	2 BC

COURIER
RATING PLATE

The information indicated on the rating plate is the vehicle identification number, body type, exterior color, interior trim and production code.

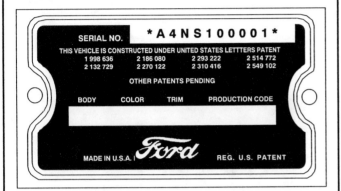

THE VEHICLE IDENTIFICATION NUMBER is a
series of letters and numbers on the rating plate. The VIN number identifies the engine, model year, assembly plant, body style and production sequence.

FIRST DIGIT: Identifies the engine

ENGINE	CODE
223 cid, 6 cyl.	A
239 cid, 8 cyl.	U

SECOND DIGIT: Identifies the model year (1954)

THIRD DIGIT: Identifies the assembly plant

ASSEMBLY PLANT	CODE
Atlanta, GA	A
Buffalo, NY	B
Chester, PA	C
Dallas, TX	D
Edgewater, NJ	E
Dearborn, MI	F
Chicago, IL	G
Highland Park	H
Kansas City, KS	K
Long Beach, CA	L
Memphis, TN	M
Norfolk, VA	N
Twin City (St. Paul), MN	P
Richmond, VA	R
Somerville, MA	S
Louisville, KY	U

FOURTH DIGIT: Identifies the body style

BODY STYLE	CODE
Sedan delivery	S

LAST SIX DIGITS: Identify the consecutive unit number

THE VEHICLE DATA appears on the line following the vehicle identification number.

THE BODY CODE indicates the body type.

BODY TYPE	CODE
Mainline	78A

THE EXTERIOR COLOR CODE indicates the paint color used on the vehicle.

COLOR	CODE
Raven Black	A
Sheridan Blue	B
Cadet Blue Metallic	C
Glacier Blue	D
Dovetone Gray	E
Highland Green Metallic	F
Killarney Green Metallic	G
Sea Haze Green	H
Lancer Maroon Metallic	J
Sandalwood Tan	L
Sandstone White	M
Cameo Coral	N
Torch Red	R
Prime	P
Special paints	SS

THE INTERIOR TRIM CODE indicates the key to the trim color and material used on the vehicle.

COLOR	CODE
Brown vinyl	AA

THE PRODUCTION CODE indicates the day of the month, the month and the car produced.

MONTH	FIRST YEAR	SECOND YEAR
January	A	N
February	B	P
March	C	Q
April	D	R
May	E	S
June	F	T
July	G	U
August	H	V
September	J	W
October	K	X
November	L	Y
December	M	Z

ENGINE SPECIFICATIONS

ENGINE CODE	NO. CYL.	CID	HORSE- POWER	COMP. RATIO	CARB
A	6	223	115	7.2:1	1 BC
U	8	239	130	7.2:1	2 BC

F-SERIES
RATING PLATE

The information indicated on the rating plate is the vehicle identification number, maximum gross vehicle weight (lbs.), wheelbase, certified net horsepower at r.p.m., transmission type, rear axle, exterior color, production code, and D.S.O. numbers.

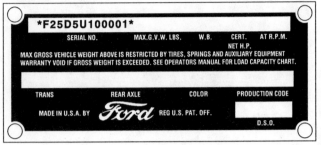

THE VEHICLE IDENTIFICATION NUMBER is a series of letters and numbers on the rating plate. The VIN number identifies the series, engine, model year, assembly plant and production sequence.

FIRST, SECOND AND THIRD DIGITS: Identify the series

SERIES	CODE
F-100	F10
F-250	F25
F-350	F35

FOURTH DIGIT: Identifies the engine

ENGINE	CODE
223 cid, 6 cyl.	D
239 cid, 8 cyl.	V
256 cid, 8 cyl.	Z

FIFTH DIGIT: Identifies the model year (1955)

SIXTH DIGIT: Identifies the assembly plant

ASSEMBLY PLANT	CODE
Atlanta, GA	A
Buffalo, NY	B
Chester, PA	C
Chicago, IL	G
Dallas, TX	D
Dearborn, MI	F
Edgewater, NJ	E
Highland Park, MI	H
Kansas City, KS	K
Long Beach, CA	L
Louisville, KY	U
Memphis, TN	M
Norfolk, VA	N
Richmond, VA	R
Somerville, MA	S
Twin City (St. Paul), MN	P

LAST SIX DIGITS: Identify the consecutive unit number

THE VEHICLE DATA appears on the two lines following the vehicle identification number.

THE MAX. G.V.W. LBS. CODE indicates the maximum gross vehicle weight in pounds.

SERIES	TYPE	GVW (LBS.)
F100	Conventional	5,000
F250	Conventional	6,900
F350	Conventional (single wheel)	7,700
F350	Conventional (dual wheel)	9,500

THE W.B. (WHEELBASE) CODE indicates the wheelbase of the vehicle in inches.

THE CERT. NET H.P. AT R.P.M. CODE indicates the certified net horsepower at the specified r.p.m.

THE TRANSMISSION CODE indicates the transmission type installed in the vehicle.

TYPE	CODE
3-Speed regular	3
3-Speed heavy duty	3HD
3-Speed automatic	AUTO
3-Speed overdrive	3OD
4-Speed synchronized	4SYN

THE REAR AXLE CODE indicates the ratio of the rear axle installed in the vehicle.

RATIO

3.92:1 ....

4.27:1 ....

4.09:1 ....

4.86:1 ....

5.14:1 ....

5.83:1 ....

THE EXTERIOR COLOR CODE indicates the paint color used on the vehicle.

COLOR	CODE
Raven Black	A
Banner Blue	B
Aquatone Blue	C
Waterfall Blue	D
Snowshoe White	E
Sea Sprite Green	G
Vermilion (Torch Red)	R
Meadow Green	U
Goldenrod Yellow	V
Prime	P
Special	SS

THE PRODUCTION CODE indicates the day of the month, the month, and the car produced.

MONTH	FIRST YEAR	SECOND YEAR
January	A	N
February	B	P
March	C	Q
April	D	R
May	E	S
June	F	T
July	G	U
August	H	V
September	J	W
October	K	X
November	L	Y
December	M	Z

THE D.S.O. CODE: Domestic Special Order

ENGINE SPECIFICATIONS

ENGINE CODE	NO. CYL.	CID	HORSE-POWER	COMP. RATIO	CARB
D	6	223	118	7.5:1	1 BC
V	8	239	106	6.8:1	2 BC
Z	8	256	140	7.5:1	2 BC

COURIER
RATING PLATE

The information indicated on the rating plate is the vehicle identification number, body type, exterior color, interior trim and production code.

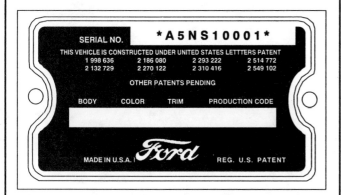

THE VEHICLE IDENTIFICATION NUMBER is a series of letters and numbers on the rating plate. The VIN number identifies the engine, model year, assembly plant, body style and production sequence.

FIRST DIGIT: Identifies the engine

ENGINE	CODE
223 cid, 6 cyl.	A
272 cid, 8 cyl.	U

SECOND DIGIT: Identifies the model year (1955)

THIRD DIGIT: Identifies the assembly plant

ASSEMBLY PLANT	CODE
Atlanta, GA	A
Buffalo, NY	B
Chester, PA	C
Dallas, TX	D
Mahwah, NJ	E
Dearborn, MI	F
Chicago, IL	G
Kansas City, KS	K
Long Beach, CA	L
Memphis, TN	M
Norfolk, VA	N
Twin City (St. Paul), MN	P
San Jose, CA	R
Somerville, MA	S
Louisville, KY	U

FOURTH DIGIT: Identifies the body style

BODY STYLE	CODE
Sedan delivery	S

LAST SIX DIGITS: Identify the consecutive unit number

THE VEHICLE DATA appears on the line following the vehicle identification number.

THE BODY CODE indicates the body type.

BODY TYPE	CODE
Sedan delivery	78A

THE EXTERIOR COLOR CODE indicates the paint color used on the vehicle.

COLOR	CODE
Raven Black	A
Banner Blue	B
Aquatone Blue	C
Waterfall Blue	D
Snowshoe White	E
Pine Tree Green	F
Sea Sprite Green	G
Neptune Green	H
Buckskin Brown	K
Regency Purple	M
Torch Red	R
Thunderbird Blue	T
Goldenrod Yellow	V
Tropical Rose	W
Prime	P
Special	S

THE INTERIOR TRIM CODE indicates the key to the trim color and material used on the vehicle.

COLOR/TYPE	CODE
Med. Copper vinyl/Dk. Copper woven plastic	AD
Dk. Copper vinyl	AE
Med. Copper vinyl/Dk. Copper western vinyl	AM

THE PRODUCTION CODE indicates the day of the month, the month, and the car produced.

MONTH	FIRST YEAR	SECOND YEAR
January	A	N
February	B	P
March	C	Q
April	D	R
May	E	S
June	F	T
July	G	U
August	H	V
September	J	W
October	K	X
November	L	Y
December	M	Z

ENGINE SPECIFICATIONS

ENGINE CODE	NO. CYL.	CID	HORSE-POWER	COMP. RATIO	CARB
A	6	223	120	7.5:1	1 BC
U	8	272	162	7.6:1	2 BC

1956 F-100 PICKUP

1956 F-100 PICKUP

1956 F-100 PICKUP

F-SERIES RATING PLATE

The information indicated on the rating plate is the vehicle identification number, maximum gross vehicle weight (lbs.), wheelbase, certified net horsepower at r.p.m., transmission type, rear axle, exterior color, production code, and D.S.O. numbers.

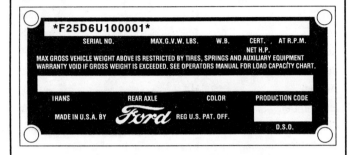

THE VEHICLE IDENTIFICATION NUMBER is a series of letters and numbers on the rating plate. The VIN identifies the series, engine, model year, assembly plant and production sequence.

FIRST, SECOND AND THIRD DIGITS: Identify the series

SERIES	CODE
F-100	F10
F-100 (LD)	F11
F-250	F25
F-250 (LD)	F26
F-350	F35

FOURTH DIGIT: Identifies the engine

ENGINE	CODE
223 cid, 6 cyl.	D
272 cid, 8 cyl. (LD)	V

FIFTH DIGIT: Identifies the model year (1956)

SIXTH DIGIT: Identifies the assembly plant

ASSEMBLY PLANT	CODE
Chicago, IL	G
Dallas, TX	D
Detroit Truck	H
Kansas City, KS	K
Long Beach, CA	L
Louisville, KY	U
Mahwah, NJ	E
Memphis, TN	M
Norfolk, VA	N
San Jose, CA	R
Twin City (St. Paul), MN	P

LAST SIX DIGITS: Identify the consecutive unit number

THE VEHICLE DATA appears on the two lines following the vehicle identification number.

THE MAX. G.V.W. LBS. CODE indicates the maximum gross vehicle weight in pounds.

SERIES	TYPE	G.V.W. (LBS.) REGULAR	LIGHT
F100	Conventional	5,000	4,000
F250	Conventional	7,400	4,900
F350	Conventional (single wheel)	8,000	7,700
F350	Conventional (dual wheel)	9,800	—

THE W.B. (WHEELBASE) CODE indicates the wheelbase of the vehicle in inches.

THE CERT. NET H.P. AT R.P.M. CODE indicates the certified net horsepower at the specified r.p.m.

THE TRANSMISSION CODE indicates the transmission type installed in the vehicle.

TYPE	CODE
3-Speed Fordomatic	AUTO
3-Speed regular	3
3-Speed med. duty	3MD
3-Speed heavy duty	3HD
3-Speed overdrive	3OD
4-Speed synchronized	4SYN

THE REAR AXLE indicates the ratio of the rear axle installed in the vehicle.

RATIO

3.73:1

3.92:1

4.56:1

4.86:1

5.14:1

5.83:1

THE EXTERIOR COLOR CODE indicates the paint color used on the vehicle.

COLOR	CODE
Raven Black	A
Dk. Blue Metallic	B
Light Blue	D
Colonial White	E
Light Green	G
Vermillion Red	R
Meadow Green	U
Yellow	M
Gray	H
Prime	P
Special	S

THE PRODUCTION CODE indicates the day of the month, the month, and the car produced. EXAMPLE: 15 J 39

MONTH	FIRST YEAR	SECOND YEAR
January	A	N
February	B	P
March	C	Q
April	D	R
May	E	S
June	F	T
July	G	U
August	H	V
September	J	W
October	K	X
November	L	Y
December	M	Z

THE D.S.O. CODE: Domestic Special Order

ENGINE SPECIFICATIONS

ENGINE CODE	NO. CYL.	CID	HORSE-POWER	COMP. RATIO	CARB
D	6	223	133	8.0:1	1 BC
V	8	272	167	8.0:1	2 BC

COURIER RATING PLATE

The information indicated on the rating plate is the vehicle identification number, body type, exterior color, interior trim and production code.

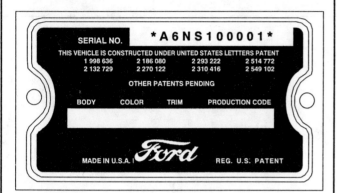

THE VEHICLE IDENTIFICATION NUMBER is a series of letters and numbers on the rating plate. The VIN number identifies the engine, model year, assembly plant, body style and production sequence.

FIRST DIGIT: Identifies the engine

ENGINE	CODE
223 cid, 6 cyl.	A
292 cid, 8 cyl.	M
272 cid, 8 cyl.	U

SECOND DIGIT: Identifies the model year (1956)

THIRD DIGIT: Identifies the assembly plant

ASSEMBLY PLANT	CODE
Atlanta, GA	A
Buffalo, NY	B
Chester, PA	C
Dallas, TX	D
Mahwah, NJ	E
Dearborn, MI	F
Chicago, IL	G
Kansas City, KS	K
Long Beach, CA	L
Memphis, TN	M
Norfolk, VA	N
Twin City (St. Paul), MN	P
San Jose, CA	R
Somerville	S
Louisville, KY	U

FOURTH DIGIT: Identifies the body style

BODY STYLE	CODE
Sedan delivery	S

LAST SIX DIGITS: Identify the consecutive unit number

THE VEHICLE DATA appears on the line following the vehicle identification number.

THE BODY CODE indicates the body type.

TYPE	CODE
Sedan delivery	78A

THE EXTERIOR COLOR CODE indicates the paint color used on the vehicle.

COLOR	CODE
Raven Black	A
Nocturne Blue	B
Bermuda Blue	C
Diamond Blue	D
Colonial White	E
Pine Ridge Green	F
Meadowmist Green	G
Platinum Gray	H
Buckskin Tan	J
Fiesta Red	K
Peacock Blue	L
Goldenglow Yellow	M
Mandarin Orange	N
Prime	P
Special	S

THE INTERIOR TRIM CODE indicates the trim color and material used on the vehicle.

COLOR	VINYL	CLOTH	LEATHER	CODE
Blue	•			AV,BM
Blue		•		H
Blue	•	•		B,BB
Green		•		J
Green	•			T
Green	•	•		C,AS,BC
Green	•	•		CA
Gray		•		L
Gray	•	•		E,BE
Brown	•			AJ,AX,BA
Brown		•		BJ
Brown	•	•		AZ
Red/White	•			V
White/Red	•			BU
White/Red	•	•		AT,BL
White/Black	•	•		AD
White/Blue	•			BV
White/Blue	•	•		AN,BH,BR
White/Blue	•	•		BS,BZ
White/Brown	•	•		BD
White/Orange	•			BG
Black/Red	•	•		AE
Black/White	•			BN
Gray/Red		•		BF

THE PRODUCTION CODE indicates the day of the month, the month, and the car produced. EXAMPLE: 15 J 39

MONTH	FIRST YEAR	SECOND YEAR
January	A	N
February	B	P
March	C	Q
April	D	R
May	E	S
June	F	T
July	G	U
August	H	V
September	J	W
October	K	X
November	L	Y
December	M	Z

ENGINE SPECIFICATIONS

ENGINE CODE	NO. CYL.	CID	HORSE-POWER	COMP. RATIO	CARB
A	6	223	137	8.0:1	1 BC
U	8	272	176	8.4:1	2 BC
M	8	292	202	8.4:1	2 BC

1957 RANCHERO

1957 F-100 STYLESIDE PICKUP

1957 RANCHERO

F-SERIES
RATING PLATE

The information indicated on the rating plate is the vehicle identification number, maximum gross vehicle weight (lbs.), wheelbase, certified net horsepower at r.p.m., transmission type, rear axle, exterior color, production code, and D.S.O. numbers.

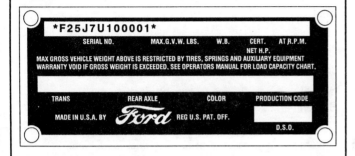

THE VEHICLE IDENTIFICATION NUMBER is a series of letters and numbers on the rating plate. The VIN number identifies the series, engine, model year, assembly plant and production sequence.

FIRST, SECOND AND THIRD DIGITS: Identify the serial code

SERIES	CODE
F-100	F10
F-100 (LD)	F11
F-250	F25
F-250 (LD)	F26
F-350	F35

FOURTH DIGIT: Identifies the engine

ENGINE	CODE
223 cid, 6 cyl.	J
272 cid, 8 cyl. (LD)	K
272 cid, 8 cyl. (HD)	L
272 cid, 8 cyl. (HD)	U

FIFTH DIGIT: Identifies the model year (1957)

SIXTH DIGIT: Identifies the assembly plant

ASSEMBLY PLANT	CODE
Chicago, IL	G
Dallas, TX	D
Detroit Truck	H
Kansas City, KS	K
Long Beach, CA	L
Louisville, KY	U
Mahwah, NJ	E
Memphis, TN	M
Norfolk, VA	N
San Jose, CA	R
Twin City (St. Paul), MN	P

LAST SIX DIGITS: Identify the consecutive unit number

THE VEHICLE DATA appears on the two lines following the vehicle identification number.

THE MAX. G.V.W. LBS. CODE indicates the maximum gross vehicle weight in pounds.

SERIES	TYPE	GVW (LBS.) REGULAR	LIGHT
F100	Conventional	5,000	4,000
F250	Conventional	7,400	4,900
F350	Conventional (single wheel)	8,000	7,700
F350	Conventional (dual wheel)	9,800	—

THE W.B. (WHEELBASE) CODE indicates the wheelbase of the vehicle in inches.

THE CERT. NET H.P. AT R.P.M. CODE indicates the certified net horsepower at the specified r.p.m.

THE TRANSMISSION CODE indicates the transmission type installed in the vehicle.

TYPE	CODE	CODE*
3-Speed Fordomatic	AUTO	C
3-Speed light duty	3LD	A
3-Speed med. duty	3MD	D
3-Speed heavy duty	3HD	E
4-Speed synchronized	4SPD	F

* New codes after 4-1-57

THE REAR AXLE indicates the ratio of the rear axle installed in the vehicle.

RATIO

3.70:1 ....
3.89:1 ....
4.11:1 ....
4.56:1 ....
4.86:1 ....
4.88:1 ....
5.14:1 ....
5.83:1 ....

THE EXTERIOR COLOR CODE indicates the paint color used on the vehicle.

COLOR	CODE
Raven Black	A
Dk. Blue Metallic	B
Starmist Blue	F
Colonial White	E
Willow Green	J
Vermillion Red	R
Meadow Green	U
Inca Gold	Y
Woodsmoke Gray	T
Prime	P
Special	S

THE PRODUCTION CODE indicates the day of the month, the month and the car produced.

MONTH	FIRST YEAR	SECOND YEAR
January	A	N
February	B	P
March	C	Q
April	D	R
May	E	S
June	F	T
July	G	U
August	H	V
September	J	W
October	K	X
November	L	Y
December	M	Z

THE D.S.O. CODE: Domestic Special Order

ENGINE SPECIFICATIONS

ENGINE CODE	NO. CYL.	CID	HORSE-POWER	COMP. RATIO	CARB
J	6	223	126	8.3:1	1 BC
K	8	272	145	8.3:1	2 BC
L	8	272	153	8.3:1	2 BC

RANCHERO
RATING PLATE

The information indicated on the rating plate is the vehicle identification number, body type, exterior color, interior trim and production code.

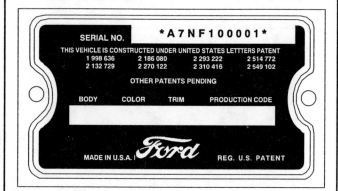

THE VEHICLE IDENTIFICATION NUMBER is a series of letters and numbers on the rating plate. The VIN number identifies the engine, model year, assembly plant, body style and production sequence.

FIRST DIGIT: Identifies the engine

ENGINE	CODE
223 cid, 6 cyl.	A
272 cid, 8 cyl.	U
292 cid, 8 cyl.	C

SECOND DIGIT: Identifies the model year (1957)

THIRD DIGIT: Identifies the assembly plant

ASSEMBLY PLANT	CODE
Atlanta, GA	A
Buffalo, NY	B
Chester, PA	C
Dallas, TX	D
Mahwah, NJ	E
Dearborn, MI	F
Chicago, IL	G
Kansas City, KS	K
Long Beach, CA	L
Memphis, TN	M
Norfolk, VA	N
Twin City (St. Paul), MN	P
San Jose, CA	R
Somerville	S
Louisville, KY	U

FOURTH DIGIT: Identifes the body style

BODY STYLE	CODE
Ranchero	F

LAST SIX DIGITS: Identify the consecutive unit number

THE VEHICLE DATA appears on the line following the vehicle identification number.

THE BODY CODE indicates the body type.

TYPE	CODE
Custom series	66A
Custom 300 series	66B

THE EXTERIOR COLOR CODE indicates the paint color used on the vehicle.

COLOR	CODE
Raven Black	A
Dresden Blue	C
Colonial White	E
Starmist Blue	F
Cumberland Green	G
Willow Green	J
Silver Mocha	K
Doeskin Tan	L
Gunmetal Gray	N
Thunderbird Bronze	Q
Woodsmoke Gray	T
Flame Red	V
Dusk Rose	X
Inca Gold	Y
Coral Sand	Z
Prime	P
Special	S

THE INTERIOR TRIM CODE indicates the key to the trim color and material used on the vehicle.

COLOR	VINYL	CLOTH	LEATHER	CODE
Silver/Gray	•	•		A
Blue		•		AR
Blue	•	•		E
Gray		•		R
Gray	•	•		F
Green		•		AS
Green	•	•		AN
Tan	•	•		AX
Tan/Brown	•	•		G
Brown		•		S
White/Tan	•	•		AZ
White/Blue	•			AB
White/Blue	•	•		H,X
White/Green	•			AC
White/Green	•	•		J
White/Gray	•	•		K
White/Brown	•	•		L
White/Black	•			AA
White/Black	•	•		U
White/Red	•			AU
White/Red	•	•		AJ
Gold/Black		•		T
Gold/Black	•	•		AT

THE PRODUCTION CODE indicates the day of the month, the month and the car produced.

MONTH	FIRST YEAR	SECOND YEAR
January	A	N
February	B	P
March	C	Q
April	D	R
May	E	S
June	F	T
July	G	U
August	H	V
September	J	W
October	K	X
November	L	Y
December	M	Z

ENGINE SPECIFICATIONS

ENGINE CODE	NO. CYL.	CID	HORSE-POWER	COMP. RATIO	CARB
A	6	223	144	8.6:1	1 BC
U	8	272	190	8.6:1	2 BC
C	8	223	212	9.1:1	2 BC

COURIER
RATING PLATE

The information indicated on the rating plate is the vehicle identification number, body type, exterior color, interior trim and production code.

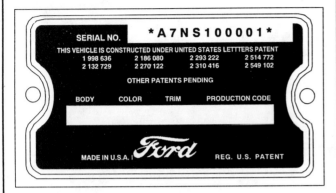

THE VEHICLE IDENTIFICATION NUMBER is a
series of letters and numbers on the rating plate. The VIN identifies the engine, model year, assembly plant, body style and production sequence.

FIRST DIGIT: Identifies the engine

ENGINE	CODE
223 cid, 6 cyl.	A
272 cid, 8 cyl.	U
292 cid, 8 cyl.	C

SECOND DIGIT: Identifies the model year (1957)

THIRD DIGIT: Identifies the assembly plant

ASSEMBLY PLANT	CODE
Atlanta, GA	A
Buffalo, NY	B
Chester, PA	C
Dallas, TX	D
Mahwah, NJ	E
Dearborn, MI	F
Chicago, IL	G
Kansas City, KS	K
Long Beach, CA	L
Memphis, TN	M
Norfolk, VA	N
Twin City (St. Paul), MN	P
San Jose, CA	R
Somerville	S
Louisville, KY	U

FOURTH DIGIT: Identifies the body style

BODY STYLE	CODE
Sedan delivery	S

LAST SIX DIGITS: Identify the consecutive unit number

THE VEHICLE DATA appears on the line following the vehicle identification number.

THE BODY CODE indicates the body type.

BODY	CODE
Sedan delivery	78A

THE EXTERIOR COLOR CODE indicates the paint color used on the vehicle.

COLOR	CODE
Raven Black	A
Dresden Blue	C
Colonial White	E
Starmist Blue	F
Cumberland Green	G
Willow Green	J
Silver Mocha	K
Doeskin Tan	L
Gunmetal Gray	N
Woodsmoke Gray	T
Flame Red	V
Inca Gold	Y
Coral Sand	Z
Prime	P
Special	S

THE INTERIOR TRIM CODE indicates the trim color and material used on the vehicle.

COLOR	VINYL	CLOTH	LEATHER	CODE
Silver	•	•		A
Blue		•		AR
Blue	•	•		E
Green		•		AS
Green	•	•		AN
Gray	•	•		F
Tan	•	•		G,AX
White/Blue	•			AB
White/Blue	•	•		H,X
White/Green	•			AC
White/Green	•	•		J
White/Gray	•	•		K
White/Brown	•	•		L
White/Tan	•	•		AZ
White/Black	•			AA
White/Black	•	•		U
White/Red	•			AU
White/Red	•	•		AJ
Gray		•		R
Brown		•		S
Gold/Black		•		T
Black/Gold	•	•		AT

THE PRODUCTION CODE indicates the day of the month, the month and the car produced. EXAMPLE: 15 J 39

MONTH	FIRST YEAR	SECOND YEAR
January	A	N
February	B	P
March	C	Q
April	D	R
May	E	S
June	F	T
July	G	U
August	H	V
September	J	W
October	K	X
November	L	Y
December	M	Z

ENGINE SPECIFICATIONS

ENGINE CODE	NO. CYL.	CID	HORSE-POWER	COMP. RATIO	CARB
A	6	223	144	8.6:1	1 BC
U	8	272	190	8.6:1	2 BC
C	8	292	212	9.10:1	2 BC

1958 RANCHERO

1958 RANCHERO

F-SERIES
RATING PLATE

The information indicated on the rating plate is the vehicle identification number, maximum gross vehicle weight (lbs.), wheelbase, certified net horsepower at r.p.m., transmission type, rear axle, exterior color, production code and D.S.O. numbers.

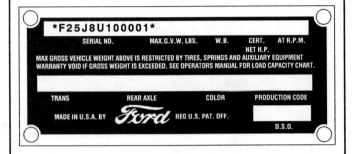

THE VEHICLE IDENTIFICATION NUMBER is a
series of letters and numbers on the rating plate. The VIN number identifies the series, engine, model year, assembly plant and production sequence.

FIRST, SECOND AND THIRD DIGITS: Identify the serial code

SERIAL NO.	CODE
F-100	F10
F-250	F25
F-350	F35

FOURTH DIGIT: Identifies the engine

ENGINE	CODE
223 cid, 6 cyl.	J
272 cid, 8 cyl. (LD)	K

FIFTH DIGIT: Identifies the model year (1958)

SIXTH DIGIT: Identifies the assembly plant

ASSEMBLY PLANT	CODE
Atlanta, GA	A
Dallas, TX	D
Mahwah, NJ	E
Chicago, IL	G
Lorain, OH	H
Kansas City, KS	K
Long Beach, CA	L
Norfolk, VA	N
Twin City (St. Paul), MN	P
San Jose, CA	R
Louisville, KY	U

LAST SIX DIGITS: Identify the consecutive unit number

THE VEHICLE DATA appears on the two lines following the vehicle identification number.

THE MAX. G.V.W. LBS. CODE indicates the maximum gross vehicle weight in pounds.

SERIES	TYPE	GVW
F-100	—	5,000
F-250	—	7,400
F-350	Pickup	7,600
F-350	Stake and Chassis cab	9,800

THE W.B. (WHEELBASE) CODE indicates the wheelbase of the vehicle in inches.

THE CERT. NET. H.P. AT R.P.M. CODE indicates the certified net horsepower at the specified r.p.m.

THE TRANSMISSION CODE indicates the transmission type installed in the vehicle.

TYPE	CODE
3-Speed standard	A
3-Speed med. duty	D
3-Speed heavy duty	E
3-Speed overdrive	B
4-Speed synchronized	F

THE REAR AXLE CODE indicates the ratio of the rear axle installed in the vehicle.

RATIO	CODE
3.70	01
3.89	02
4.11	03
4.56	04
4.86	05
4.88	06
5.14	07
5.83	08

FRONT AXLE:

RATIO	CODE
3.92	A
4.55	B

THE EXTERIOR COLOR CODE indicates the paint color used on the vehicle.

COLOR	CODE
Raven Black	A
Dark Blue	B
Colonial White	E
Med. Gray Poly.	H
Light Blue	L
Light Green	N
Dark Green	U
Red	R
Goldenrod Yellow	X

THE D.S.O. CODE: Domestic Special Order (No codes available)

ENGINE SPECIFICATIONS

ENGINE CODE	NO. CYL.	CID	HORSE-POWER	COMP. RATIO	CARB
J	6	223	126	8.3:1	1 BC
K	8	272	145	8.3:1	2 BC

RANCHERO
RATING PLATE

The information indicated on the rating plate is the vehicle identification number, body type, exterior color, interior trim and production code.

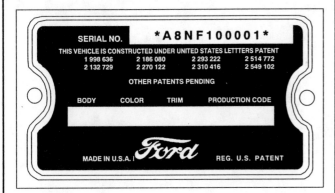

THE VEHICLE IDENTIFICATION NUMBER is a series of letters and numbers on the rating plate. The VIN number identifies the engine, model year, assembly plant, body style and production sequence.

FIRST DIGIT: Identifies the engine

ENGINE	CODE
223 cid, 6 cyl.	A
292 cid, 8 cyl.	C
352 cid, 8 cyl.	H

SECOND DIGIT: Identifies the model year (1958)

THIRD DIGIT: Identifies the assembly plant

ASSEMBLY PLANT	CODE
Atlanta, GA	A
Chester	C
Dallas, TX	D
Mahwah, NJ	E
Dearborn, MI	F
Chicago, IL	G
Lorain, OH	H
Los Angeles, CA	J
Kansas City, KS	K
Long Beach, CA	L
Memphis, TN	M
Norfolk, VA	N
Twin Cities, MN	P
San Jose, CA	R
Allen Park, MI (Pilot Plant)	S
Metuchen, NJ	T
Louisville, KY	U
Wayne, MI	W
Wixom, MI	Y
St. Louis, MO	Z

FOURTH DIGIT: Identifies the body style

BODY STYLE	CODE
Ranchero	F

LAST SIX DIGITS: Identify the consecutive unit number

THE VEHICLE DATA appears on the line following the vehicle identification number.

THE BODY CODE indicates the body type.

BODY TYPE	CODE
Custom series	66A
Custom 300 series	66B

THE EXTERIOR COLOR CODE indicates the paint color used on the vehicle.

COLOR	CODE
Raven Black	A
Desert Beige	C
Colonial White	E
Silvertone Green	F
Gunmetal Gray	H
Bali Bronze	J
Azure Blue	L
Seaspray Green	N
Torch Red	R
Silvertone Blue	T

THE INTERIOR TRIM CODE indicates the trim color and material used on the vehicle.

COLOR	VINYL	CLOTH	LEATHER	CODE
Blue	•			U
Blue		•		J
Blue	•	•		C,AC,CZ
Green	•			V
Green		•		K
Green	•	•		D,DZ
Brown		•		L
Beige	•	•		AB
Beige/Brown	•	•		F,AH
White	•			AL
White	•	•		AK
White/Red	•			Y
White/Red	•	•		S
White/Black	•			Z
White/Black	•	•		G
White/Palomino	•			X
White/Palomino	•	•		H
Black/Gold		•		R
Black/Gold	•	•		AM

THE PRODUCTION CODE indicates the day of the month, the month, and the car produced. EXAMPLE: 15 J 39

MONTH	FIRST YEAR	SECOND YEAR
January	A	N
February	B	P
March	C	Q
April	D	R
May	E	S
June	F	T
July	G	U
August	H	V
September	J	W
October	K	X
November	L	Y
December	M	Z

ENGINE SPECIFICATIONS

ENGINE CODE	NO. CYL.	CID	HORSE-POWER	COMP. RATIO	CARB
A	6	223	145	8.6:1	1 BC
C	8	292	205	9.1:1	2 BC
H	8	352	300	9.6:1	4 BC

COURIER
RATING PLATE

The information indicated on the rating plate is the vehicle identification number, body type, exterior color, interior trim and production code.

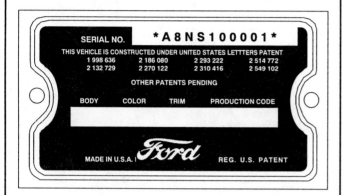

THE VEHICLE IDENTIFICATION NUMBER is a series of letters and numbers on the rating plate. The VIN number identifies the engine, model year, assembly plant, body style and production sequence.

FIRST DIGIT: Identifies the engine

ENGINE	CODE
223, cid, 6 cyl.	A
292 cid, 8 cyl.	C
352 cid, 8 cyl.	H

SECOND DIGIT: Identifies the model year (1958)

THIRD DIGIT: Identifies the assembly plant

ASSEMBLY PLANT	CODE
Atlanta, GA	A
Chester, PA	C
Dallas, TX	D
Mahwah, NJ	E
Dearborn, MI	F
Chicago, IL	G
Lorain, OH	H
Los Angeles, CA	J
Kansas City, KS	K
Long Beach, CA	L
Memphis, TN	M
Norfolk, VA	N
Twin Cities, MN	P
San Jose, CA	R
Allen Park, MI (Pilot Plant)	S
Metuchen, NJ	T
Louisville, KY	U
Wayne, MI	W
Wixom, MI	Y
St. Louis, MO	Z

FOURTH DIGIT: Identifies the body style

BODY STYLE	CODE
Sedan delivery	S

LAST SIX DIGITS: Identify the consecutive unit number

THE VEHICLE DATA appears on the line following the vehicle identification number.

THE BODY CODE indicates the body type.

BODY TYPE	CODE
Sedan delivery	78A

THE EXTERIOR COLOR CODE indicates the paint color used on the vehicle.

COLOR	CODE
Raven Black	A
Dark Blue	B
Colonial White	E
Gunmetal Gray	H
Light Blue	L
Light Green	N
Red	R
Dark Green	U
Goldenrod Yellow	X

THE INTERIOR TRIM CODE indicates the trim color and material used on the vehicle.

COLOR	CODE
Med. Blue shantung vinyl	39
Med. Brown shantung vinyl	47

THE PRODUCTION CODE indicates the day of the month, the month, and the car produced. EXAMPLE: 15 J 39

MONTH	FIRST YEAR	SECOND YEAR
January	A	N
February	B	P
March	C	Q
April	D	R
May	E	S
June	F	T
July	G	U
August	H	V
September	J	W
October	K	X
November	L	Y
December	M	Z

ENGINE SPECIFICATIONS

ENGINE CODE	NO. CYL.	CID	HORSE-POWER	COMP. RATIO	CARB
A	6	223	145	8.6:1	1 BC
C	8	292	205	9.1:1	2 BC
H	8	352	300	9.6:1	4 BC

F-SERIES
RATING PLATE

The information indicated on the rating plate is the vehicle identification number, maximum gross vehicle weight (lbs.), wheelbase, certified net horsepower at r.p.m., transmission type, rear axle, exterior color, production code, and D.S.O. numbers.

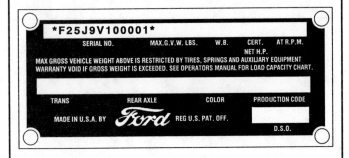

THE VEHICLE IDENTIFICATION NUMBER is a series of letters and numbers on the rating plate. The VIN number identifies the series, engine, model year, assembly plant and production sequence.

FIRST, SECOND AND THIRD DIGITS: Identify the series

SERIES	CODE
F-100	F10
F-100 (4x4)	F11
F-250	F25
F-250 (4x4)	F26
F-350	F35

FOURTH DIGIT: Identifies the engine

ENGINE	CODE
292 cid, 8 cyl.	C
292 cid, 8 cyl.	D
223 cid, 6 cyl.	J

FIFTH DIGIT: Identifies the model year (1959)

SIXTH DIGIT: Identifies the assembly plant

ASSEMBLY PLANT	CODE
Atlanta, GA	A
Chicago, IL	G
Dallas, TX	D
Kansas City, KS	K
Long Beach, CA	L
Lorain, OH	H
Louisville, KY	U
Mahwah, NJ	E
Norfolk, VA	N
Allen Park, MI (Pilot Plant)	S
San Jose, CA	R
Twin City (St. Paul), MN	P

LAST SIX DIGITS: Identify the consecutive unit number

THE VEHICLE DATA appears on the two lines following the vehicle identification number.

THE MAX. G.V.W. LBS. CODE indicates the maximum gross vehicle weight in pounds.

SERIES	TYPE	GVW (LBS.)
F-100		5,000
F-250		7,400
F-350	Pickup	7,600

THE W.B. (WHEELBASE) CODE indicates the wheelbase of the vehicle in inches.

THE CERT. NET H.P. AT R.P.M. CODE indicates the certified net horsepower at the specified r.p.m.

THE EXTERIOR COLOR CODE indicates the paint color used on the vehicle.

COLOR	CODE
Raven Black	A
Wedgewood Blue	C
Indian Turquoise	D
Colonial White	E
April Green	G
Red	R
Meadow Green	U
Academy Blue	V
Goldenrod Yellow	X

THE TRANSMISSION CODE indicates the transmission type installed in the vehicle.

TYPE	CODE
3-Speed standard	A
3-Speed (MD)	D
3-Speed (HD)	E
3-Speed w/overdrive	B
4-Speed synchronized	F

THE REAR AXLE CODE indicates the ratio of the rear axle installed in the vehicle.

RATIO	CODE
3.70	01
3.89	02
4.11	03
4.56	04
4.86	05
4.88	06
5.14	07
5.83	08
3.73	23
3.92	24
4.88	25
4.56	26

FRONT AXLE:

RATIO	CODE
3.92	A
4.55	B

THE D.S.O. CODE: Domestic Special Order (Codes not available)

ENGINE SPECIFICATIONS

ENGINE CODE	NO. CYL.	CID	HORSE-POWER	COMP. RATIO	CARB
J	6	223	126	8.1:1	1 BC
C	8	292	158	7.9:1	2 BC
D	8	292	160	7.6:1	4 BC

RANCHERO RATING PLATE

The information indicated on the rating plate is the vehicle identification number, body type, exterior color, interior trim and production code.

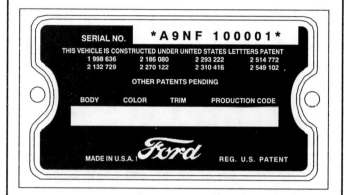

THE VEHICLE IDENTIFICATION NUMBER is a series of letters and numbers on the rating plate. The VIN number identifies the engine, model year, assembly plant, body and production sequence.

FIRST DIGIT: Identifies the engine

ENGINE	CODE
223 cid, 6 cyl.	A
292 cid, 8 cyl.	C
352 cid, 8 cyl.	H

SECOND DIGIT: Identifies the model year (1959)

THIRD DIGIT: Identifies the assembly plant

ASSEMBLY PLANT	CODE
Atlanta, GA	A
Chester, PA	C
Dallas, TX	D
Mahwah, NJ	E
Dearborn, MI	F
Chicago, IL	G
Lorain, OH	H
Los Angeles, CA	J
Kansas City, KS	K
Long Beach, CA	L
Memphis, TN	M
Norfolk, VA	N
Twin Cities, MN	P
San Jose, CA	R
Allen Park, MI (Pilot Plant)	S
Metuchen, NJ	T
Louisville, KY	U
Wayne, MI	W
Wixom, MI	Y
St. Louis, MO	Z

FOURTH DIGIT: Identifies the body style

BODY	CODE
Ranchero	F

LAST SIX DIGITS: Identify the consecutive unit number

THE VEHICLE DATA appears on the line following the vehicle identification number.

THE BODY CODE indicates the body type.

BODY TYPE	CODE
Custom Ranchero	66C

THE EXTERIOR COLOR CODE indicates the paint color used on the vehicle.

COLOR	CODE
Raven Black	A
Wedgewood Blue	C
Colonial White	E
Fawn Tan	F
April Green	G
Surf Blue	J
Sherwood Green	Q
Torch Red	R
Inca Gold	Y
Prime	P
Surf Blue	L
Tahitian Bronze	K
Gunsmoke Gray	N

THE INTERIOR TRIM CODE indicates the trim color and material used on the vehicle.

COLOR	VINYL	CLOTH	LEATHER	CODE
Blue	•			39,49
Blue		•		301
Blue	•	•		02,07,25
Blue	•	•		30,024,302
Med. Blue	•			59
Med. Brown	•			60
Green	•			40
Green		•		311
Green	•	•		03,08,31
Green	•	•		45,032,312
Bronze	•			47
Bronze		•		331
Bronze	•	•		09,33,331
Bronze	•	•		332
Gray		•		121,323
Gray	•	•		10,12
Turquoise	•			51
Turquoise		•		111
Turquoise	•	•		11,34,341
Fawn	•			26
Gunsmoke	•	•		29

COLOR	VINYL	CLOTH	LEATHER	CODE
Silver	•	•		32,296
White	•			58
White	•	•		37,46,57
White	•	•		371
White/Red	•	•		571
Gold	•			52
Gold	•	•		38,381
Geranium	•			53
Geranium	•	•		48
Red	•			55
Red	•	•		56

THE PRODUCTION CODE indicates the day of the month, the month, and the car produced. EXAMPLE: 15 J 39

MONTH	FIRST YEAR	SECOND YEAR
January	A	N
February	B	P
March	C	Q
April	D	R
May	E	S
June	F	T
July	G	U
August	H	V
September	J	W
October	K	X
November	L	Y
December	M	Z

ENGINE SPECIFICATIONS:

ENGINE CODE	NO. CYL.	CID	HORSE-POWER	COMP. RATIO	CARB
A	6	223	145	8.4:1	1 BC
C	8	292	200	8.8:1	2 BC
H	8	352	300	9.6:1	4 BC

COURIER
RATING PLATE

The information indicated on the rating plate is the vehicle identification number, body type, exterior color, interior trim and production code.

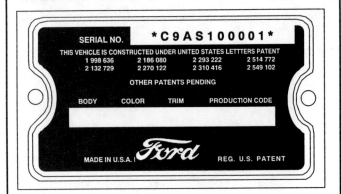

THE VEHICLE IDENTIFICATION NUMBER is a series of letters and numbers on the rating plate. The VIN number identifies the engine, model year, assembly plant, body style and production sequence.

FIRST DIGIT: Identifies the engine

ENGINE	CODE
223 cid, 6 cyl.	A
292 cid, 8 cyl.	C
332 cid, 8 cyl.	B
352 cid, 8 cyl.	H

SECOND DIGIT: Identifies the model year (1959)

THIRD DIGIT: Identifies the assembly plant

ASSEMBLY PLANT	CODE
Atlanta, GA	A
Chester, PA	C
Dallas, TX	D
Mahwah, NJ	E
Dearborn, MI	F
Chicago, IL	G
Lorain, OH	H
Los Angeles, CA	J
Kansas City, KS	K
Long Beach, CA	L
Memphis, TN	M
Norfolk, VA	N
Twin Cities, MN	P
San Jose, CA	R
Allen Park, MI (Pilot Plant)	S
Metuchen, NJ	T
Louisville, KY	U
Wayne, MI	W
Wixom, MI	Y
St. Louis, MO	Z

FOURTH DIGIT: Identifies the body style

BODY	CODE
Sedan delivery	S

LAST SIX DIGITS: Identify the consecutive unit number

THE VEHICLE DATA appears on the line following the vehicle identification number.

THE BODY CODE indicates the body type.

BODY TYPE	CODE
Sedan delivery	78A

THE EXTERIOR COLOR CODE indicates the paint color used on the vehicle.

COLOR	CODE
Raven Black	A
Wedgewood Blue	C
Colonial White	E
Fawn Tan	F
April Green	G
Tahitian Bronze	H
Sherwood Green	Q
Torch Red	R
Inca Gold	Y
Surf Blue	L
Gunsmoke Gray	N

THE INTERIOR TRIM CODE indicates the trim color and material used on the vehicle.

COLOR/TYPE	CODE
Med. Blue shantung vinyl	39
Med. Brown shantung vinyl	47

THE PRODUCTION CODE indicates the day of the month, the month, and the car produced. EXAMPLE: 15 J 39

MONTH	FIRST YEAR	SECOND YEAR
January	A	N
February	B	P
March	C	Q
April	D	R
May	E	S
June	F	T
July	G	U
August	H	V
September	J	W
October	K	X
November	L	Y
December	M	Z

ENGINE SPECIFICATIONS:

ENGINE CODE	NO. CYL.	CID	HORSE-POWER	COMP. RATIO	CARB
A	6	223	145	8.4:1	1 BC
C	8	292	200	8.8:1	2 BC
B	8	332	225	8.9:1	2 BC
H	8	352	300	9.6:1	4 BC

F-SERIES
RATING PLATE

The information indicated on the rating plate is the vehicle identification number, wheelbase, exterior color, model type, date manufactured, transmission type, rear axle, maximum gross vehicle weight (lbs.), certified net horsepower, r.p.m., and D.S.O. numbers.

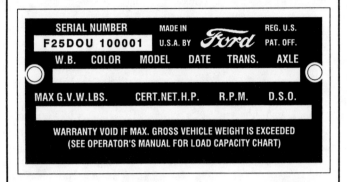

THE VEHICLE IDENTIFICATION NUMBER is a series of letters and numbers on the rating plate. The VIN number identifies the series, engine, model year, assembly plant and production sequence.

FIRST, SECOND AND THIRD DIGITS: Identify the series

SERIES	CODE
F-100	F10
F-100 (4x4)	F11
F-250	F25
F-250 (4x4)	F26
F-350	F35

FOURTH DIGIT: Identifies the engine

ENGINE	CODE
292 cid, 8 cyl. (MD)	C
223 cid, 6 cyl.	J

FIFTH DIGIT: Identifies the model year (1960)

SIXTH DIGIT: Identifies the assembly plant

ASSEMBLY PLANT	CODE
Atlanta, GA	A
Dallas, TX	D
Mahwah, NJ	E
Chicago, IL	G
Lorain, OH	H
Kansas City, KS	K
Norfolk, VA	N
Twin City (St. Paul), MN	P
San Jose, CA	R
Louisville, KY	U

LAST SIX DIGITS: Identify the consecutive unit number

THE VEHICLE DATA appears on the two lines following the vehicle identification number.

THE W.B. (WHEELBASE) CODE indicates the wheelbase of the vehicle in inches. No special codes are used.

THE EXTERIOR COLOR CODE indicates the paint color used on the vehicle.

COLOR	CODE
Raven Black	A
Corinthian White	M
Academy Blue	V
Sky Mist Blue	F
Dark Green	L
Adriatic Green	W
Monte Carlo Red	J
Goldenrod Yellow	X
Turquoise	B

THE MODEL CODE indicates the model type and the G.V.W. (lbs.) information.

MODEL	GVW
F-100	5,000
F-101	4,000
F-102	5,000
F-103	5,600
F-104	4,000
F-105	5,600
F-250	7,400
F-251	4,900
F-252	7,400
F-253	4,900
F-350	9,800
F-351	7,700

THE DATE CODE indicates the date the vehicle was manufactued. A number signifying the date precedes the month code letter. A second year code letter will be used if the model exceeds 12 months.

MONTH	FIRST YEAR	SECOND YEAR
January	A	N
February	B	P
March	C	Q
April	D	R
May	E	S
June	F	T
July	G	U
August	H	V
September	J	W
October	K	X
November	L	Y
December	M	Z

THE TRANSMISSION CODE indicates the transmission type installed in the vehicle.

TYPE	CODE
3-Speed standard	A
3-Speed overdrive	B
Fordomatic	C
3-Speed MD Warner T-89C	D
3-Speed HD Warner T-87E	E
4-Speed Warner T-98A	F
HD Cruise-O-Matic	G

THE REAR AXLE CODE indicates the ratio of the rear axle installed in the vehicle. The first two digits identify the rear axle and the third digit identifies the front axle.

RATIO	CODE
3.73	OA
3.92	OB
4.56	OD
4.88	OF
3.70	01
3.89	02
4.11	03
4.56	04
4.86	05
4.88	06
5.14	07
5.83	08

FRONT AXLE

RATIO	CODE
3.92 (4x4)	A
4.55 (4x4)	B
6,000 lb.	C
7,000 lb.	D
9,000 lb.	E

THE MAX. G.V.W. LBS. CODE indicates the maximum gross vehicle weight in pounds.

THE CERTIFIED NET H.P. AT R.P.M. CODE indicates the certified net horsepower at the specified r.p.m.

ENGINE SPECIFICATIONS:

ENGINE CODE	NO. CYL.	CID	HORSE-POWER	COMP. RATIO	CARB
J	6	223	126	8.1:1	1 BC
C	8	292	146	7.9:1	2 BC

RANCHERO RATING PLATE

The information indicated on the rating plate is the vehicle identification number, body type, exterior color, interior trim, date manufactured, transmission type and rear axle.

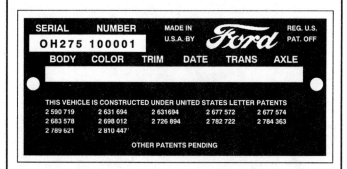

THE VEHICLE IDENTIFICATION NUMBER is a

series of letters and numbers on the rating plate. The VIN number identifies the model year, assembly plant, model, engine and production sequence.

FIRST DIGIT: Identifies the model year (1960)

SECOND DIGIT: Identifies the assembly plant

ASSEMBLY PLANT	CODE
Atlanta, GA	A
Lorain, OH	H
Kansas City, KS	K
San Jose, CA	R
Allen Park, MI (Pilot Plant)	S
Metuchen, NJ	T

THIRD AND FOURTH DIGITS: Identify the model

MODEL	CODE
Ranchero	27
Sedan delivery	29

FIFTH DIGIT: Identifies the engine

ENGINE	CODE
144 cid, 6 cyl.	S
144 cid, 6 cyl. (LC)	D
170 cid, 6 cyl.	U
170 cid, 6 cyl. (LC)	E

LAST SIX DIGITS: Identify the consecutive unit number

THE VEHICLE DATA appears on the line following the vehicle identification number.

THE BODY CODE indicates the body type.

BODY	CODE
Ranchero	66A
Sedan delivery	78A

THE EXTERIOR COLOR CODE indicates the paint color used on the vehicle. Two-tone paint codes use the same symbols as the single colors except that two symbols are used. The first symbol is the lower color, the second symbol is the upper color.

COLOR	CODE
Raven Black	A
Belmont Blue	E
Sky Mist Blue	F
Beachwood Brown	H
Monte Carlo Red	J
Sultana Turquoise	K
Corinthian White	M
Meadowvale Green	T
Adriatic Green	W
Platinum	Z

THE INTERIOR TRIM CODE indicates the trim color and material used on the vehicle.

FIRST DIGIT	MATERIAL	SECOND DIGIT	COLOR
1	Vinyl/Cloth	0	Silver/White
2	Vinyl/Cloth	1	Gray
4	Vinyl/Cloth	2	Blue/Lt. Blue & Med. Blue
5	Vinyl	3	Green
6	Vinyl/Cloth	4	Brown/Tan/Beige
7	Vinyl	5	Red/Red & White
8	Futura Vinyl	6	Black/Black & White
—	—	7	Turquoise

THE DATE CODE indicates the date the vehicle was manufactured. A number signifying the date precedes the month code letter. a second year code letter will be used if the mdoel exceeds 12 months.

MONTH	FIRST YEAR	SECOND YEAR
January	A	N
February	B	P
March	C	Q
April	D	R
May	E	S
June	F	T
July	G	U
August	H	V
September	J	W
October	K	X
November	L	Y
December	M	Z

THE TRANSMISSION CODE indicates the transmission type installed in the vehicle.

TYPE	CODE
Manual shift	1
Fordomatic	3

THE REAR AXLE CODE indicates the ratio of the rear axle installed in the vehicle.

RATIO	CODE
3.56	1
3.89	2

ENGINE SPECIFICATIONS:

ENGINE CODE	NO. CYL.	CID	HORSE-POWER	COMP. RATIO	CARB
S	6	144	90	8.7:1	1 BC
U	6	170	101	8.7:1	1 BC

COURIER RATING PLATE

The information indicated on the rating plate is the vehicle identification number, body type, exterior color, interior trim, date manufactured, transmission type and rear axle.

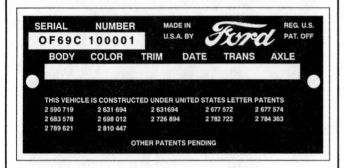

THE VEHICLE IDENTIFICATION NUMBER is a series of letters and numbers on the rating plate. The VIN number identifies the model year, assembly plant, model, engine and production sequence.

FIRST DIGIT: Identifies the model year (1960)

SECOND DIGIT: Identifies the assembly plant

ASSEMBLY PLANT	CODE
Atlanta, GA	A
Chester	C
Dallas, TX	D
Mahwah, NJ	E
Dearborn, MI	F
Chicago, IL	G
Lorain, OH	H
Los Angeles, CA	J
Kansas City, KS	K
Long Beach, CA	L
Memphis, TN	M
Norfolk, VA	N
Twin Cities, MN	P
San Jose, CA	R
Allen Park, MI (Pilot Plant)	S
Metuchen, NJ	T
Louisville, KY	U
Wayne, MI	W
Wixom, MI	Y
St. Louis, MO	Z

THIRD AND FOURTH DIGITS: Identify the model

MODEL	CODE
Sedan delivery	69

FIFTH DIGIT: Identifies the engine

ENGINE	CODE
223 cid, 6 cyl.	V
292 cid, 8 cyl., 2 BC	W
352 cid, 8 cyl., 2 BC	X
352 cid, 8 cyl., 4 BC	Y
352 cid, 8 cyl., 4 BC (LC)	G
392 cid, 8 cyl., 2 BC (LC)	T

LAST SIX DIGITS: Identify the consecutive unit number

THE VEHICLE DATA appears on the line following the vehicle identification number.

THE BODY CODE indicates the body type.

BODY	CODE
Sedan delivery	59E

THE EXTERIOR COLOR CODE indicates the paint color used on the vehicle.

COLOR	CODE
Raven Black	A
Aqua Marine	C
Belmont Blue	E
Sky Mist Blue	F
Yosemite Yellow	G
Beachwood Brown	H
Monte Carlo Red	J
Sultana Turquoise	K
Corinthian White	M
Prime	P
Orchid Gray	Q
Special	S
Meadowvale Green	T
Adriatic Green	W
Platinum	Z

THE INTERIOR TRIM CODE indicates the trim color and material used on the vehicle.

COLOR/TYPE	CODE
Blue vinyl/broadcloth	32
Blue vinyl	42
Blue vinyl/woven plastic	62
Green vinyl/broadcloth	33
Green vinyl	43
Green vinyl/woven plastic	63
Gray vinyl/broadcloth	31
Beige vinyl/broadcloth	34
Beige vinyl	44
Turquoise vinyl/broadcloth	37
Turquoise vinyl	47
Turquoise vinyl/woven plastic	67
Black vinyl/broadcloth	36
Black vinyl	46
Yellow vinyl/broadcloth	38
Yellow vinyl	48
Red vinyl/broadcloth	35
Red vinyl	45
Lavender vinyl/broadcloth	39
Lavender vinyl	49

THE PRODUCTION CODE indicates the day of the month, the month, and the car produced. EXAMPLE: 15 J 39

MONTH	FIRST YEAR	SECOND YEAR
January	A	N
February	B	P
March	C	Q
April	D	R
May	E	S
June	F	T
July	G	U
August	H	V
September	J	W
October	K	X
November	L	Y
December	M	Z

ENGINE SPECIFICATIONS:

ENGINE CODE	NO. CYL.	CID	HORSE-POWER	COMP. RATIO	CARB
V	6	223	145	8.4:1	1 BC
W	8	292	185	8.8:1	2 BC
X	8	352	220	8.9:1	2 BC
Y	8	352	300	9.6:1	4 BC

1961 FALCON SEDAN DELIVERY

1961 F-100 PICKUP

1961 FALCON RANCHERO

1961 F-100 PICKUP

1961 F-100 PICKUP

F-SERIES RATING PLATE

The information indicated on the rating plate is the vehicle identification number, wheelbase, exterior color, model type, date of production, transmission type, rear axle, maximum gross vehicle weight (lbs.), certified net horsepower, r.p.m., and D.S.O. numbers. Beginning in 1961 a consecutive unit numbering system was implemented which identifies the year and month of production by its unit number and not by a date code.

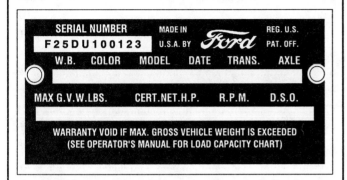

THE VEHICLE IDENTIFICATION NUMBER is a series of letters and numbers on the rating plate. The VIN number identifies the series, engine, assembly plant and production sequence.

FIRST, SECOND AND THIRD DIGITS: Identify the series

SERIES	CODE
F-100	F10
F-100	F11
F-250	F25
F-250	F26
F-350	F35

FOURTH DIGIT: Identifies the engine

ENGINE	CODE
292 cid, 8 cyl. (MD)	C
292 cid, 8 cyl. (HD)	D
223 cid, 6 cyl.	J

FIFTH DIGIT: Identifies the assembly plant

ASSEMBLY PLANT	CODE
Dallas, TX	D
Mahwah, NJ	E
Chicago, IL	G
Lorain, OH	H
Kansas City, KS	K
Norfolk, VA	N
Twin Cities, MN	P
San Jose, CA	R
Louisville, KY	U

LAST SIX DIGITS: Identifies the consecutive unit number

CALENDAR YEAR 1961	NUMBERS
October	100,001 thru 109,999
November	110,000 thru 119,999
December	120,000 thru 129,999
January	130,000 thru 139,999
February	140,000 thru 149,999
March	150,000 thru 159,999
April	160,000 thru 169,999
May	170,000 thru 179,999
June	180,000 thru 189,999
July	190,000 thru 199,999

THE VEHICLE DATA appears on the two lines following the vehicle identification number.

THE W.B. (WHEELBASE) CODE indicates the wheelbase of the vehicle in inches. No special codes are used.

THE EXTERIOR COLOR CODE indicates the paint color used on the vehicle.

COLOR	CODE
Raven Black	A
Corinthian White	M
Academy Blue	V
Dark Green	L
Monte Carlo Red	J
Goldenrod Yellow	X
Turquoise	B
Light Blue	D
Light Green	S

THE MODEL CODE indicates the model type and the gross vehicle weight (lbs.) information.

F-100 SERIES

MODEL	GVW
F-100	5,000
F-101	4,000
F-102	5,000

F-100 (4X4) SERIES

MODEL	GVW
F-113	5,600
F-114	4,000
F-115	5,600

F-250 SERIES

MODEL	GVW
F-250	7,400
F-251	4,900

F-250 (4X4) SERIES

MODEL	GVW
F-262	6,600
F-263	4,900
F-264	7,400

F-350 SERIES

MODEL	GVW
F-350	9,800
F-351	7,600

THE DATE CODE indicates the date the vehicle was manufactured. No date codes were used. It is identified by the consecutive unit number.

THE TRANSMISSION CODE indicates the transmission type installed in the vehicle.

TYPE	CODE
3-Speed standard	A
3-Speed overdrive	B
Fordomatic	C
3-Speed M/D Warner T-89C	D
3-Speed H/D Warner T-87E	E
4-Speed Warner T-98A	F
H/D Cruise-O-Matic	G

THE REAR AXLE CODE indicates the ratio of the rear axle installed in the vehicle.

REAR AXLE

RATIO	CODE
3.73	A1
3.92	A2
4.56	B4
4.88	B6
3.50	01
4.00	02
3.22	10
3.70	11
3.89	12
4.11	13
4.56	24
4.86	25
4.88	26
5.14	27
5.83	28

FRONT AXLE

RATIO	CODE
3.92 (4-wheel drive)	A
4.55 (4-wheel drive)	B
4.55 (4-wheel drive)	D

THE MAX. G.V.W. LBS. CODE indicates the maximum gross vehicle weight in pounds.

THE CERT. NET H.P. CODE indicates the certified net horsepower at the specified r.p.m.

THE R.P.M. CODE indicates the specified r.p.m. required to develop the certified net horsepower.

THE D.S.O. CODE: If the vehicle is built on a Direct Special Order the complete order number will be reflected under the D.S.O. space including the District Code Number.

ENGINE SPECIFICATIONS

ENGINE CODE	NO. CYL.	CID	HORSE-POWER	COMP. RATIO	CARB
J	6	223	114	8.4:1	1 BC
C	8	292	135	8.0:1	2 BC
D	8	292	153	8.0:1	4 BC
N	8	302	163	7.5:1	4 BC

RANCHERO & SEDAN DELIVERY RATING PLATE

The information indicated on the rating plate is the vehicle identification number, body type, exterior color, interior trim, date manufactured, transmission type and rear axle.

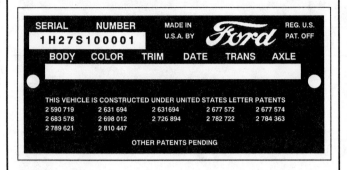

THE VEHICLE IDENTIFICATION NUMBER is a
series of letters and numbers on the rating plate. The VIN number identifies the model year, assembly plant, model, engine and production sequence.

FIRST DIGIT: Identifies the model year (1961)

SECOND DIGIT: Identifies the assembly plant

ASSEMBLY PLANT	CODE
Atlanta, GA	A
Lorain, OH	H
Kansas City, KS	K
San Jose, CA	R
Allen Park, MI (Pilot Plant)	S
Metuchen, NJ	T

THIRD AND FOURTH DIGITS: Identify the model

The model code number shows the product line series in the first digit. The second digit shows the body type: an odd number shows a 2-door model, while an even number shows a 4-door model.

MODEL	CODE
Ranchero	27
Sedan delivery	29

FIFTH DIGIT: Identifies the engine

ENGINE	CODE
144 cid, 6 cyl.	S
170 cid, 6 cyl.	U

LAST SIX DIGITS: Identify the consecutive unit number

THE VEHICLE DATA appears on the line following the vehicle identification number.

THE BODY CODE indicates the body type.

BODY	CODE
Ranchero	66A
Sedan delivery	78A

THE EXTERIOR COLOR CODE indicates the paint
color used on the vehicle.

COLOR	CODE
Black	A
Light Turquoise	C
Light Blue	D
Medium Green Metallic	E
Dark Blue Metallic	H
Red	J
Bronze Metallic	K
White	M
Light Gray Metallic	Q
Med. Blue Metallic	R
Light Green	S
Turquoise Metallic	W

THE INTERIOR TRIM CODE indicates the trim color
and the material used on the vehicle. The trim code includes 2 digits.

FIRST DIGIT	MATERIAL	SECOND DIGIT	COLOR
1	Vinyl/bodycloth	0	Silver or White
2	Vinyl/bodycloth	1	Gray
4	Vinyl/		Blue or Lt. Blue
	tweed bodycloth	2	& Med. Blue
5	Vinyl	3	Green
6	Vinyl/woven plastic		
	or bodycloth	4	Brown, Tan or Beige
7	Vinyl and vinyl	5	Red or Red & White
8	Futura vinyl	6	Black or Black & White
—	—	7	Turquoise

THE DATE CODE indicates the date the vehicle was manufactured. A number signifying the date precedes the month code letter. A second year code letter will be used if the mdoel exceeds 12 months.

MONTH	FIRST YEAR	SECOND YEAR
January	A	N
February	B	P
March	C	Q
April	D	R
May	E	S
June	F	T
July	G	U
August	H	V
September	J	W
October	K	X
November	L	Y
December	M	Z

THE TRANSMISSION CODE indicates the transmission type installed in the vehicle.

TYPE	CODE
Manual shift	1
Fordomatic	3

THE REAR AXLE CODE indicates the ratio of the rear axle installed in the vehicle.

RATIO	CODE
3.10	3
4.00	4
3.20	5
3.50	J

ENGINE SPECIFICATIONS

ENGINE CODE	NO. CYL.	CID	HORSE-POWER	COMP. RATIO	CARB
S	6	144	90	8.7:1	
U	6	170	101	8.7:1	

1962 F-100 PICKUP

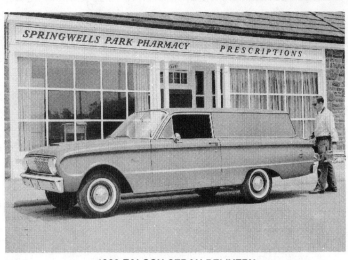

1962 FALCON SEDAN DELIVERY

F-SERIES RATING PLATE

The information indicated on the rating plate is the vehicle identification number, wheelbase, exterior color, model type, transmission type, rear axle, maximum gross vehicle weight (lbs.), certified net horsepower, r.p.m., and D.S.O. numbers.

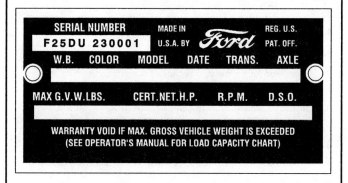

THE VEHICLE IDENTIFICATION NUMBER is a series of letters and numbers on the rating plate. The VIN number identifies the series, engine, assembly plant and production sequence.

FIRST, SECOND AND THIRD DIGITS: Identify the series

SERIES	CODE
F-100 (4x2)	F10
F-100 (4x4)	F11
F-250 (4x2)	F25
F-250 (4x4)	F26
F-350 (4x2)	F35

FOURTH DIGIT: Identifies the engine

ENGINE	CODE
292 cid, 8 cyl. (MD)	C
292 cid, 8 cyl. (HD)	D
223 cid, 6 cyl.	J

FIFTH DIGIT: Identifies the assembly plant

ASSEMBLY PLANT	CODE
Dallas, TX	D
Mahwah, NJ	E
Chicago, IL	G
Lorain, OH	H
Kansas City, KS	K
Norfolk, VA	N
Twin Cities, MN	P
San Jose, CA	R
Allen Park, MI (Pilot Plant)	S
Louisville, KY	U

LAST SIX DIGITS: Identifies the consecutive unit number

CALENDAR YEAR 1962	NUMBERS
August	205,000 thru 209,999
September	210,000 thru 219,999
October	220,000 thru 229,999
November	230,000 thru 239,999
December	240,000 thru 249,999
January	250,000 thru 259,999
February	260,000 thru 269,999
March	270,000 thru 279,999
April	280,000 thru 289,999
May	290,999 thru 299,999
June	300,000 thru 309,999
July	310,000 thru 319,999
August	320,000 thru 329,999
September	330,000 thru 339,999

THE VEHICLE DATA appears on the two lines following the vehicle identification number.

THE W.B. (WHEELBASE) CODE indicates the wheelbase of the vehicle in inches. No special codes are used.

THE EXTERIOR COLOR CODE indicates the paint color used on the vehicle.

COLOR	CODE
Raven Black	A
White	C
Corinthian White	M
Baffin Blue	F
Academy Blue	V
Dark Green	L
Rangoon Red	J
Chrome Yellow	G
Goldenrod Yellow	X
Caribbean Turquoise	B
Sandshell Beige	T

THE MODEL CODE indicates the model type and the G.V.W. (lbs.) information.

F-100 SERIES

MODEL	GVW
F-100	5,000
F-101	4,000
F-102	5,000

F-100 (4X4) SERIES

	CODE
F-113	5,600
F-114	4,000
F-115	5,600

F-250 SERIES

MODEL	GVW
F-250	7,400
F-251	4,900

F-250 (4X4) SERIES

MODEL	GVW
F-262	6,600
F-263	4,900
F-264	7,400

F-350 SERIES

MODEL	GVW
F-350	9,800
F-351	7,600

THE BODY CODE indicates the body type.

BODY TYPE	CODE
Platform	80
Chassis w/cab	81
Stake	86

THE TRANSMISSION CODE indicates the transmission type installed in the vehicle.

DESCRIPTION	CODE
3-Speed standard	A
3-Speed overdrive	B
Fordomatic	C
3-Speed M/D Warner T-89C	D
3-Speed H/D Warner T-87E	E
4-Speed Warner T-98A	F
H/D Cruise-O-Matic	G

THE REAR AXLE CODE indicates the ratio of the rear axle installed in the vehicle. The first two digits identify the rear axle and the third digit identifies the front axle.

REAR AXLE

RATIO	CODE
3.73	A1
3.92	A2
4.56	B4
4.88	B6
3.50	01
4.00	02
3.22	10
3.70	11
3.89	12
4.11	13
4.88	22
5.13	23
4.56	24
4.88	26
5.87	29

FRONT AXLE

RATIO	CODE
3.92 (4-wheel drive)	A
4.55 (4-wheel drive)	B
6,000 lb.	C
4.55 (4-wheel drive)	D
7,000 lb.	E
9,000 lb.	F
11,000 lb.	G

THE MAX. G.V.W. LBS. CODE indicates the maximum gross vehicle weight in pounds.

THE CERT. NET. H.P. CODE indicates the certified net horsepower at the specified r.p.m..

THE R.P.M. CODE indicates the specified r.p.m. required to develop the certified net horsepower.

THE D.S.O. CODE: Units built on a Domestic Special Order, Foreign Special Order, or other special orders will have the complete order number in this space. If the unit is a regular production unit, this space will be blank.

ENGINE SPECIFICATIONS:

ENGINE CODE	NO. CYL.	CID	HORSE-POWER	COMP. RATIO	CARB
J	6	223	114	8.4:1	1 BC
C	8	292	135	8.0:1	2 BC
D	8	292	153	8.0:1	4 BC

RANCHERO & SEDAN DELIVERY RATING PLATE

The information indicated on the rating plate is the vehicle identification number, body type, exterior color, interior trim, date manufactured, D.S.O., rear axle and transmission type.

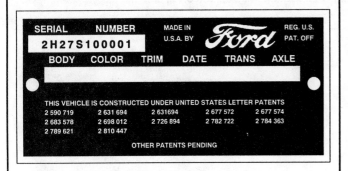

THE VEHICLE IDENTIFICATION NUMBER is a series of letters and numbers on the rating plate. The VIN number identifies the model year, assembly plant, model, engine and production sequence.

FIRST DIGIT: Identifies the model year (1962)

SECOND DIGIT: Identifies the assembly plant

ASSEMBLY PLANT	CODE
Atlanta, GA	A
Lorain, OH	H
Kansas City	K
San Jose, CA	R
Allen Park, MI (Pilot Plant)	S
Metuchen, NJ	T

THIRD AND FOURTH DIGITS: Identify the model

The model code number shows the product line series in the first digit. The second digit shows the body type: an odd number shows a 2-door model, while an even number shows a 4-door model.

MODEL	CODE
Ranchero	27
Sedan delivery	29

FIFTH DIGIT: Identifies the engine

ENGINE	CODE
144 cid, 6 cyl.	S
170 cid, 6-cyl.	U

LAST SIX DIGITS: Identify the consecutive unit number

THE VEHICLE DATA appears on the line preceding the vehicle identification number.

THE BODY CODE indicates the body type.

TYPE	CODE
Ranchero	66A
Sedan delivery	78A

THE EXTERIOR COLOR CODE indicates the paint color used on the vehicle.

COLOR	CODE
Black	A
Med. Turquoise Metallic	D
Med. Blue Metallic	E
Light Blue	F
Dk. Blue Metallic	H
Red	J
White	M
Med. Green Metallic	P
Light Gray Metallic	Q
Yellow	R
Honey Beige	T
Beige Metallic	Z

THE INTERIOR TRIM CODE indicates the trim color and material used on the vehicle. The trim code includes 2 digits.

FIRST DIGIT	MATERIAL	SECOND DIGIT	COLOR
1	Vinyl/bodycloth	0	Silver or White
2	Vinyl/bodycloth	1	Gray
4	Vinyl/ tweed bodycloth	2	Blue or Lt. Blue & Med. Blue
5	Vinyl	3	Green
6	Vinyl/woven plastic or bodycloth	4	Brown, Tan or Beige
7	Vinyl and vinyl	5	Red or Red & White
8	Futura vinyl	6	Black or Black & White

THE DATE CODE indicates the date the vehicle was manufactured. A number signifying the date precedes the month code letter. A second year code letter will be used if the model exceeds 12 months.

MONTH	FIRST YEAR	SECOND YEAR
January	A	N
February	B	P
March	C	Q
April	D	R
May	E	S
June	F	T
July	G	U
August	H	V
September	J	W
October	K	X
November	L	Y
December	M	Z

THE TRANSMISSION CODE indicates the transmission type installed in the vehicle.

TRANSMISSION	CODE
Manual-Shift	1
Fordomatic	3

THE REAR AXLE CODE indicates the ratio of the rear axle installed in the vehicle.

RATIO	CODE
3.10	2
3.50	5
4.00	9

THE D.S.O. CODE: Units built on a Domestic Special Order, Foreign Special Order, or other special orders will have the complete order number in this space. If the unit is a regular production unit, this space will be blank.

ENGINE SPECIFICATIONS:

ENGINE CODE	NO. CYL.	CID	HORSE- POWER	COMP. RATIO	CARB
S	6	144	85	8.7:1	1 BC
U	6	170	101	8.7:1	1 BC

1963 F-100 PICKUP

1963 RANCHERO

F-SERIES
RATING PLATE

The information indicated on the rating plate is the vehicle identification number, wheelbase, exterior color, model type, body type, transmission type, rear axle, maximum gross vehicle weight (lbs.), certified net horsepower, r.p.m. and D.S.O. numbers.

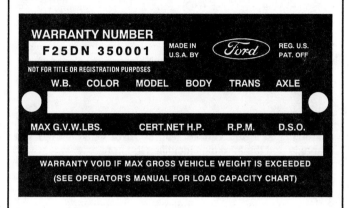

THE VEHICLE IDENTIFICATION NUMBER is a series of letters and numbers on the rating plate. The VIN number identifies the series, engine, assembly plant and production sequence.

FIRST, SECOND AND THIRD DIGITS: Identify the series

SERIES	CODE
F-100 (4x2)	F10
F-100 (4x4)	F11
F-250 (4x2)	F25
F-250 (4x4)	F26
F-350 (4x2)	F35

THIRD DIGIT: Identifies the engine

ENGINE	CODE
262 cid	B
292 cid	C
292 cid	D
302 cid	N

FIFTH DIGIT: Identifies the assembly plant

ASSEMBLY PLANT	CODE
Dallas, TX	D
Mahwah, NJ	E
Chicago, IL	G
Lorain, OH	H
Kansas City, KS	K
Norfolk, VA	N
Twin City (St. Paul), MN	P
San Jose, CA	R
Allen Park, MI (Pilot Plant)	S

LAST SIX DIGITS: Identify the consecutive unit number

MONTH	NUMBERS
August	325,000 thru 329,999
September	330,000 thru 339,999
October	340,000 thru 349,999
November	350,000 thru 359,999
December	360,000 thru 369,999
January	370,000 thru 379,999
February	380,000 thru 389,999
March	390,000 thru 399,999
April	400,000 thru 409,999
May	410,000 thru 419,999
June	420,000 thru 429,999
July	430,000 thru 439,999
August	440,000 thru 449,999
September	450,000 thru 459,999

THE VEHICLE DATA appears on the two lines following the vehicle identification number.

THE W.B. (WHEELBASE) CODE indicates the wheelbase in inches. No special codes are used.

THE EXTERIOR COLOR CODE indicates the paint color used on the vehicle.

COLOR	CODE
Raven Black	A
Carribean Turquoise	B
Chrome Yellow	C
Glacier Blue	Y
White	G
Rangoon Red	J
Holly Green	L
Corinthian White	M
Sandshell Beige	T
Academy Blue	V
Driftwood	K

THE MODEL CODE indicates the model type and the G.V.W. (lbs.) information

MODEL	GVW
F-100	5,000
F-101	4,000
F-102	5,000
F-110 (4x4)	5,600
F-111 (4x4)	4,000
F-112	5,600
F-250	7,400
F-251	4,900
F-260 (4x4)	6,600
F-261 (4x4)	4,900
F-262 (4x4)	7,400
F-350	9,800
F-351	7,800

THE BODY CODE indicates the body type.

BODY TYPE	CODE
Platform	80
Chassis w/cab	81
Stake	86

THE TRANSMISSION CODE indicates the transmission type installed in the vehicle.

TYPE	CODE
*3-Speed Ford standard duty	A
*3-Speed Ford w/Warner T86 overdrive	B
*3-Speed Warner T89-C (MD)	D
*3-Speed Warner T87-E (HD)	E
*3-Speed HD Cruise O-Matic	G
4-Speed Warner T98-A	F

* Transmissions not used at Louisville assembly plant

THE REAR AXLE CODE indicates the ratio of the rear axle installed in the vehicle. The first two digits identify the rear axle and the third digit identifies the front axle.

RATIO	CODE
3.70	11
3.89	12
4.11	13
5.83	21
4.88	22
5.13	23
4.56	24
4.88	26
3.73	A1
3.92	A2
4.10	A5
4.56	B4
4.88	B6

THE MAX. G.V.W. LBS. CODE indicates the recommended maximum gross vehicle weight in pounds.

THE CERT. NET. H.P. AT R.P.M. CODE indicates the certified net horsepower at the specified r.p.m..

THE D.S.O. CODE: Trucks built to Domestic Special Order have the order number and the District Code number of the district which ordered the unit stamped in this space. If the truck is a regular production unit, only the District Code will appear.

DISTRICT	CODE
Boston	11
Buffalo	12
New York	13
Pittsburgh	14
Newark	15
Atlanta	21
Charlotte	22
Philadelphia	23
Jacksonville	24
Richmond	25
Washington	26
Buffalo	31
Cleveland	32
Detroit	33
Indianapolis	34
Lansing	35
Louisville	36
Chicago	41
Fargo	42
Rockford	43
Twin Cities	44
Davenport	45
Denver	51
Des Moines	52
Kansas City	53
Omaha	54
St. Louis	55
Dallas	61
Houston	62
Memphis	63
New Orleans	64
Oklahoma City	65
Los Angeles	71
San Jose	72
Salt Lake City	73
Seattle	74
Ford of Canada	81
Government	83
Home Office Reserve	84
American Red Cross	85
Diplomatic Service Comm.	86
Transportation Service	89
Export	90's

ENGINE SPECIFICATIONS:

ENGINE CODE	NO. CYL.	CID	HORSE-POWER	COMP. RATIO	CARB
B	6	262	132	7.9:1	1 BC
C	8	292	135	8.0:1	2 BC
D	8	292	153	8.0:1	4 BC
N	8	302	159	7.5:1	2 BC

RANCHERO & SEDAN DELIVERY RATING PLATE

The information indicated on the rating plate is the vehicle identification number, body type, exterior color, interior trim, date manufactured, D.S.O., rear axle and transmission type.

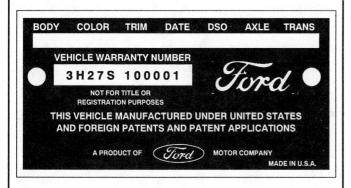

THE VEHICLE IDENTIFICATION NUMBER is a
series of letters and numbers on the rating plate. The VIN number identifies the model year, assembly plant, model, engine and production sequence.

FIRST DIGIT: Identifies the model year (1963)

SECOND DIGIT: Identifies the assembly plant

ASSEMBLY PLANT	CODE
Atlanta, GA	A
Lorain, OH	H
Los Angeles, CA	J
San Jose, CA	R
Allen Park, MI (Pilot Plant)	S
Metuchen, NJ	T

THIRD AND FOURTH DIGITS: Identify the model

MODEL	SERIES
Ranchero	27
Sedan delivery	29

FIFTH DIGIT: Identifies the engine

ENGINE	CODE
144 cid, 6 cyl.	S
170 cid, 6 cyl.	U
260 cid, 8 cyl.	F

LAST SIX DIGITS: Identify the consecutive unit numbers

THE VEHICLE DATA appears on the line above the vehicle identification number.

THE BODY CODE indicates the body type.

BODY	CODE
2-Door Ranchero	66A
2-Door deluxe Ranchero	66B
2-Door sedan delivery	78A
2-Door deluxe sedan delivery	78B

THE EXTERIOR COLOR CODE indicates the paint
color used on the vehicle. A single letter code designates a solid body color and two letters denote a two-tone; the first letter, the lower color and the second letter, the upper color.

COLOR	CODE
Raven Black	A
Ming Green Metallic	D
Viking Blue Metallic	E
Oxford Blue Metallic	H
Champagne	I
Rangoon Red	J
Corinthian White	M
Silver Moss Metallic	P
Sandshell Beige	T
Rose Beige	W
Heritage Burgundy	X
Glacier Blue	Y

THE INTERIOR TRIM CODE indicates the trim color and material used on the vehicle. A two digit number indicates the type of trim color. If, due to unavailability or other difficulties in production, a particular trim set is not intended for service (minor from intended trim), the warranty plate code will be followed with a numerical designation. For example: 52-1, 52-2. If the deviation trim set is serviced directly, the warranty plate code will bear an alphabetical suffix. For example: 52-A, 52-B.

COLOR	VINYL	CLOTH	LEATHER	CODE
Lt./Med. Blue Met.	°	°		22,92
Pearl Beige	•	°		24
Red	•	•		25,95
Lt./Med. Turquoise Met.	•	•		27
Lt. Gold Met.	•	•		28,18,98
Lt./Med. Blue	°			52
Red	°			55,75,85
Black	°			56,86
Lt./Med. Turquoise	•			57,77,87
Lt. Gold	°			58,78,88
Lt./Med. Blue Met.	°			72,82

THE DATE CODE indicates the date the vehicle was manufactured. A number signifying the date precedes the month code letter. A second year code letter will be used if the model exceeds 12 months.

MONTH	FIRST YEAR	SECOND YEAR
January	A	N
February	B	P
March	C	Q
April	D	R
May	E	S
June	F	T
July	G	U
August	H	V
September	J	W
October	K	X
November	L	Y
December	M	Z

THE TRANSMISSION CODE indicates the transmission type installed in the vehicle.

TYPE	CODE
3-Speed manual	1
2-Speed automatic	3
4-Speed manual	5

THE REAR AXLE CODE indicates the ratio of the rear axle installed in the vehicle.

RATIO	CODE
3.10:1	2
3.20:1	3
3.50:1	5
4.00:1	9

THE D.S.O. CODE: Units built on a Domestic Special Order, Foreign Special Order, or other special orders will have the complete order number in this space. Also to appear in this space is the two digit code number of the District which ordered the unit. If the unit is a regular production unit, only the District code number will appear.

DISTRICT	CODE
Boston	11
Buffalo	12
New York	13
Pittsburgh	14
Newark	15
Atlanta	21
Charlotte	22
Philadelphia	23
Jacksonville	24
Richmond	25
Washington	26
Cincinnati	31
Cleveland	32
Detroit	33
Indianapolis	34
Lansing	35
Louisville	36
Chicago	41
Fargo	42
Rockford	43
Twin Cities	44
Davenport	45
Denver	51
Des Moines	52
Kansas City	53
Omaha	54
St. Louis	55
Dallas	61
Houston	62
Memphis	63
New Orleans	64
Oklahoma City	65
Los Angeles	71
San Jose	72
Salt Lake City	73
Seattle	74
Ford of Canada	81
Government	83
Home Office Reserve	84
American Red Cross	85
Transportation Servies	89
Export	90's

ENGINE SPECIFICATIONS:

ENGINE CODE	NO. CYL.	CID	HORSE-POWER	COMP. RATIO	CARB
S	6	144	85	8.7:1	1 BC
U	6	170	101	8.7:1	1 BC
F	8	260	164	8.7:1	2 BC

1964 F-100 PICKUP

1964 FALCON RANCHERO

F-SERIES
RATING PLATE

The information indicated on the rating plate is the vehicle identification number, wheelbase, exterior color, model type, body type, transmission type, rear axle, maximum gross vehicle weight (lbs.), certified net horsepower, r.p.m. and D.S.O. numbers.

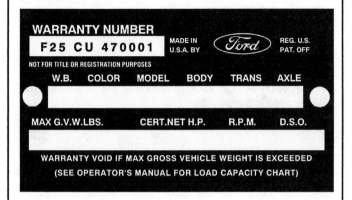

THE VEHICLE IDENTIFICATION NUMBER is a
series of letters and numbers on the rating plate. The VIN number identifies the series, engine, assembly plant and production sequence.

FIRST, SECOND AND THIRD DIGITS: Identify the series

SERIES	CODE
F-100 (4x2)	F10
F-100 (4x4)	F11
F-250 (4x2)	F25
F-250 (4x4)	F26
F-350 (4x2)	F35

THIRD DIGIT: Identifies the engine

ENGINE	CODE
223 cid	J
262 cid	B
292 cid	C

FOURTH DIGIT: Identifies the assembly plant

ASSEMBLY PLANT	CODE
Atlanta, GA	A
Dallas, TX	D
Mahwah, NJ	E
Chicago, IL	G
Lorain, OH	H
Kansas City, KS	K
Michigan Truck	L
Norfolk, VA	N
Twin Cities, MN	P
San Jose, CA	R
Allen Park, MI (Pilot Plant)	S
Louisville, KY	U

THE LAST SIX DIGITS: Identifies the consecutive unit number

MONTH	NUMBERS
August	445,000 thru 449,999
September	450,000 thru 459,999
October	460,000 thru 469,999
November	470,000 thru 479,999
December	480,000 thru 489,999
January	490,000 thru 499,999
February	500,000 thru 509,999
March	510,000 thru 519,999
April	520,000 thru 529,999
May	530,000 thru 539,999
June	540,000 thru 549,999
July	550,000 thru 559,999

THE VEHICLE DATA appears on the two lines following the vehicle identification number.

THE W.B. (WHEELBASE) CODE indicates the wheelbase in inches. No special codes are used.

THE EXTERIOR COLOR CODE indicates the paint color used on the vehicle.

COLOR	CODE
Raven Black	A
Caribbean Turquoise	B
Pure White	C
Chrome Yellow	G
Rangoon Red	J
Bengal Tan	K
Holly Green	L
Wimbledon White	M
Mint Green	S
Navajo Beige	T
Academy Blue	V
Skylight Blue	Y

THE MODEL CODE indicates the model type and the gross vehicle weight (lbs.) information.

MODEL	GVW
F-100	5,000
F-101	4,200
F-102*	5,000
F-110 (4x4)	5,600
F-111 (4x4)	4,900
F-112* (4x4)	5,600
F-250	7,500
F-251	4,800
F-260 (4x4)	6,800
F-261 (4x4)	4,900
F-262 (4x4)	7,700
F-350	10,000
F-351	8,000

* Reference Pennsylvania registration data

BODY TYPE	CODE
Platform	80
Chassis w/cab	81
Stake	86

THE TRANSMISSION CODE indicates the transmission type installed in the vehicle.

TYPE	CODE
3-Speed standard duty	A
3-Speed Ford w/Warner T86 overdrive	B
3-Speed Warner T89-C (MD)	D
3-Speed Warner T87-E (HD)	E
3-Speed HD Cruise-O-Matic	G
4-Speed Warner T98-A	F

THE REAR AXLE CODE indicates the ratio of the rear axle installed in the vehicle.

RATIO	CODE
3.70	11
3.89	12
4.11	13
4.88	22
5.13	23
4.56	24
4.10	25
4.88	26
5.87	29
3.73	A1
3.92	A2
4.11	A3
4.56	A4
4.10	A5
4.56	B4

THE MAX. G.V.W. LBS. CODE indicates the maximum gross vehicle weight in pounds.

THE CERT. NET. H.P. AT R.P.M. CODE indicates the certified net horsepower at specified rpm.

THE D.S.O. CODE: Trucks built to Domestic Special Order have the order number and the District Code Number of the district which ordered the unit stamped in this space. If the truck is a regular production unit, only the District Code will appear.

DISTRICT	CODE
Boston	11
Buffalo	12
New York	13
Pittsburgh	14
Newark	15
Atlanta	21
Charlotte	22
Philadelphia	23
Jacksonville	24
Richmond	25
Washington	26
Buffalo	31
Cleveland	32
Detroit	33
Indianapolis	34
Lansing	35
Louisville	36
Chicago	41
Fargo	42
Rockford	43
Twin Cities	44
Davenport	45
Denver	51
Des Moines	52
Kansas City	53
Omaha	54
St. Louis	55
Dallas	61
Houston	62
Memphis	63
New Orleans	64
Oklahoma City	65
Los Angeles	71
San Jose	72
Salt Lake City	73
Seattle	74
Ford of Canada	81
Government	83
Home Office Reserve	84
American Red Cross	85
Diplomatic Service Comm.	86
Transportation Service	89
Export	90's

ENGINE SPECIFICATIONS:

ENGINE CODE	NO. CYL.	CID	HORSE-POWER	COMP. RATIO	CARB
J	6	223	135	8.1:1	1 BC
B	6	262	152	7.9:1	1 BC
C	8	292	160	8.0:1	2 BC

RANCHERO & SEDAN DELIVERY RATING PLATE

The information indicated on the rating plate is the vehicle identification number, body type, exterior color, interior trim, date manufactured, D.S.O., rear axle and transmission type.

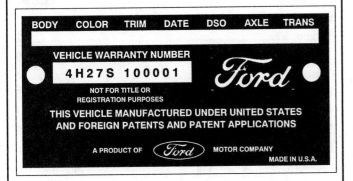

BODY	COLOR	TRIM	DATE	DSO	AXLE	TRANS

VEHICLE WARRANTY NUMBER

4H27S 100001 *Ford*

NOT FOR TITLE OR
REGISTRATION PURPOSES

THIS VEHICLE MANUFACTURED UNDER UNITED STATES
AND FOREIGN PATENTS AND PATENT APPLICATIONS

A PRODUCT OF *Ford* MOTOR COMPANY

MADE IN U.S.A.

THE VEHICLE IDENTIFICATION NUMBER is a series of letters and numbers on the rating plate. The VIN number identifies the model year, assembly plant, model type, engine and production sequence.

FIRST DIGIT: Identifies the model year (1964)

SECOND DIGIT: Identifies the assembly plant

ASSEMBLY PLANT	CODE
Atlanta, GA	A
Lorain, OH	H
Kansas City, KS	K
San Jose, CA	R
Allen Park, MI (Pilot Plant)	S
Metuchen, NJ	T

THIRD AND FOURTH DIGITS: Identify the model

MODEL	SERIES
Ranchero	27
Sedan delivery	29

FIFTH DIGIT: Identifies the engine

ENGINE	CODE
144 cid, 6 cyl.	S
170 cid, 6 cyl.	U
200 cid, 6 cyl.	T
260 cid, 8 cyl.	*F

*Low compression

LAST SIX DIGITS: Identify the consecutive unit number

THE VEHICLE DATA appears on the line preceding the vehicle identification number.

THE MODEL CODE indicates the model type.

TYPE	CODE
2-Door standard	66A
2-Door deluxe	66B
2-Door standard sedan delivery	78A
2-Door deluxe sedan delivery	78B

THE EXTERIOR COLOR CODE indicates the paint color used on the vehicle. A single letter code designates a solid body color and two letters denote a two-tone; the first letter, the lower color and the second letter, the upper color.

COLOR	CODE
Raven Black	A
Dynasty Green	D
Guardsman Blue	F
Prairie Tan	G
Rangoon Red	J
Silvermore Gray	K
Wimbledon White	M
Vintage Burgundy	X
Skylight Blue	Y
Chantilly Beige	Z

THE INTERIOR TRIM CODE indicates the key to the trim color and material used on the vehicle.

COLOR	VINYL	CLOTH	LEATHER	CODE
Black	•			66,86
Lt. Blue Met.	•	•		12,22
Med./Lt. Blue Met.	•			62,82
Lt. Beige	•			44,64
Lt. Beige Met.	•	•		14,24
Red	•			65,85
Red	•	•		15,25
Lt. Turquoise Met.	•	•		27
Med./Lt. Turquoise Met.	•			67,87
Med. Palomino	•			69,89

THE TRANSMISSION CODE indicates the transmission type installed in the vehicle.

TYPE	CODE
3-Speed manual	1
2-Speed automatic	3
Dual range	4
4-Speed manual	5

THE REAR AXLE CODE indicates the ratio of the rear axle installed in the vehicle. A number designates a conventional axle, while a letter designates an Equa-Lock differential.

RATIO	CODE
3.10:1	2
3.20:1	3
3.25:1	4
3.50:1	5
4.00:1	9
3.10:1	B
3.20:1	C
3.25:1	D
3.50:1	E
4.00:1	I

THE DATE CODE indicates the date the vehicle was manufactured. A number signifying the date precedes the month code letter. A second year code letter will be used if the model exceeds 12 months.

MONTH	FIRST YEAR	SECOND YEAR
January	A	N
February	B	P
March	C	Q
April	D	R
May	E	S
June	F	T
July	G	U
August	H	V
September	J	W
October	K	X
November	L	Y
December	M	Z

THE D.S.O. CODE: Units built on a Domestic Special Order, Foreign Special Order, or other special orders will have the complete order number in this space. Also to appear in this space is the two-digit code number of the District which ordered the unit. If the unit is a regular production unit, only the District code number will appear.

DISTRICT	CODE
Boston	11
Buffalo	12
New York	13
Pittsburgh	14
Newark	15
Altanta	21
Charlotte	22
Philadelphia	23
Jacksonville	24
Richmond	25
Washington	26
Cincinnati	31
Cleveland	32
Detroit	33
Indianpolis	34
Lansing	35
Louisville	36
Chicago	41
Fargo	42
Rockford	43
Twin Cities	44
Davenport	45
Denver	51
Des Moines	52
Kansas City	53
Omaha	54
St. Louis	55
Dallas	61
Houston	62
Memphis	63
New Orleans	64
Oklahoma City	65
Los Angeles	71
San Jose	72
Salt Lake City	73
Seattle	74
Ford of Canada	81
Government	83
Home Office Reserve	84
American Red Cross	85
Transportation Services	89
Export	90's

ENGINE SPECIFICATIONS

ENGINE CODE	NO. CYL.	CID	HORSE-POWER	COMP. RATIO	CARB
S	6	144	85	8.7:1	1 BC
U	6	170	101	8.7:1	1 BC
T	6	200	116	8.7:1	1 BC
F	8	260	164	8.7:1	2 BC

1965 F-100 STYLESIDE

1965 F-250 PICKUP

1965 RANCHERO

F-SERIES
RATING PLATE

The information indicated on the rating plate is the vehicle identification number, wheelbase, exterior color, model type, body type, transmission type, rear axle, maximum gross vehicle weight (lbs.), certified net horsepower, r.p.m. and D.S.O. numbers.

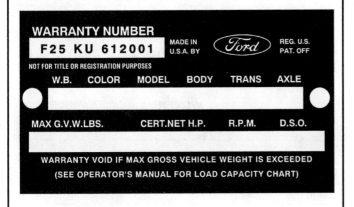

THE VEHICLE IDENTIFICATION NUMBER is a

series of letters and numbers on the rating plate. The VIN number identifies the series, engine, assembly plant and production sequence.

FIRST, SECOND AND THIRD DIGITS: Identify the series

SERIES	CODE
F-100 (4x2)	F10
F-100 (4x4)	F11
F-250 (4x2)	F25
F-250 (4x4)	F26
F-350 (4x2)	F35

FOURTH DIGIT: Identifies the engine

ENGINE	CODE
300 cid, 6 cyl.	B
352 cid, 8 cyl.	D
240 cid, 6 cyl.	J

FIFTH DIGIT: Identifies the assembly plant

ASSEMBLY PLANT	CODE
Atlanta, GA	A
Dallas, TX	D
Mahwah, NJ	E
Chicago, IL	G
Lorain, OH	H
Los Angeles, CA	J
Kansas City, KS	K
Michigan Truck	L
Norfolk, VA	N
Twin Cities, MN	P
San Jose, CA	R
Allen Park, MI (Pilot Plant)	S
Metuchen, NJ	T
Louisville, KY	U
Wayne, MI	W
Wixom, MI	Y
St. Louis, MO	Z

LAST SIX DIGITS: Identify the consecutive unit number

CALENDAR YEAR 1964	NUMBERS
August	580,000 thru 587,999
September	588,000 thru 599,999
October	600,000 thru 611,999
November	612,000 thru 623,999
December	624,000 thru 635,999
January	636,000 thru 647,999
February	648,000 thru 659,999
March	660,000 thru 671,999
April	672,000 thru 683,999
May	684,000 thru 695,999
June	696,000 thru 707,999
July	708,000 thru 719,999
August	720,000 thru 731,999

THE VEHICLE DATA appears on the two lines following the vehicle identification number.

THE W.B. (WHEELBASE) CODE indicates the wheelbase in inches.

THE EXTERIOR COLOR CODE indicates the paint color used on the vehicle.

COLOR	CODE
Black	A
Turquoise	B
Special White (RPO)	C
Chrome Yellow	G
Red	J
Tan	K
Dk. Green	L
White	M
Lt. Peacock	O
Palomino Metallic	P
Yellow	V
Med. Blue	W

THE MODEL CODE indicates the model type and the gross vehicle weight (lbs.) information.

MODEL	GVW
F-100	5,000
F-101	4,200
F-102*	5,000
F-110 (4x4)	5,600
F-111 (4x4)	4,900
F-112* (4x4)	5,600
P-100	4,300
P-101	5,000
F-250	7,500
F-251	4,800
F-260 (4x4)	6,800
F-261 (4x4)	4,900
F-262 (4x4)	7,700
F-350	10,000
F-351	8,000

* Reference Pennsylvania registration data

THE BODY CODE indicates the body type and the interior trim color and material used on the vehicle.

BODY	CODE
Conventional cab	81
Cowl and chassis	84
Cowl and windshield	85

THE INTERIOR TRIM CODE indicates the key to the trim color and material used on the vehicle.

COLOR	VINYL	WOVEN PLASTIC	LEATHER	CODE
Blue	•			2
Blue	•	•		B,K
Green	•			3
Beige	•			4
Beige	•	•		D,M
Red	•			5
Red	•	•		E,N
Black	•			6,O
Gray	•			J
Gray	•	•		A
Green	•	•		C
Green	•	•		L

THE TRANSMISSION CODE indicates the transmission type installed in the vehicle.

TYPE	CODE
3-Speed Ford standard duty	A
3-Speed Ford w/Warner T86 overdrive	B
3-Speed Warner T89-C (MD)	D
3-Speed Warner T87-E (HD)	E
4-Speed Warner T98-A	F
3-Speed HD Cruise-O-Matic	G
4-Speed New Process 435	N

THE REAR AXLE CODE indicates the ratio of the rear axle installed in the vehicle.

RATIO	CODE
3.25	07
3.50	08
3.70	09
4.11	10
4.88	22
5.13	23
4.10	24
4.56	25
4.88	26
5.87	29
3.54	A8
3.54	A9
4.10	B4
4.56	B5
4.88	B6
3.31	C1
3.73	C2
3.92	C3
4.09	C4
4.10	C5

THE MAX. G.V.W. LBS. CODE indicates the maximum gross vehicle weight in pounds.

THE CERT. NET. H.P. AT R.P.M. CODE indicates the certified net horsepower at specified rpm.

THE D.S.O. CODE: If the vehicle is built on a Direct Special Order, Foreign Special Order, or L.P.O., the complete order number will be reflected under the DSO space, including the District Code Number.

DISTRICT	CODE
Boston	11
Buffalo	12
New York	13
Pittsburgh	14
Newark	15
Atlanta	21
Charlotte	22
Philadelphia	23
Jacksonville	24
Richmond	25
Washington	26
Buffalo	31
Cleveland	32
Detroit	33
Indianapolis	34
Lansing	35
Louisville	36
Chicago	41
Fargo	42
Rockford	43
Twin Cities	44
Davenport	45
Denver	51
Des Moines	52
Kansas City	53
Omaha	54
St. Louis	55
Dallas	61
Houston	62
Memphis	63
New Orleans	64
Oklahoma City	65
Los Angeles	71
San Jose	72
Salt Lake City	73
Seattle	74
Ford of Canada	81
Government	83
Home Office Reserve	84
American Red Cross	85
Transportation Services	89
Export	90's

ENGINE SPECIFICATIONS:

ENGINE CODE	NO. CYL.	CID	HORSE-POWER	COMP. RATIO	CARB
J	6	240	150	8.75:1	1 BC
B	6	300	170	8.0:1	1 BC
D	8	352	208	8.9:1	2 BC

RANCHERO & SEDAN DELIVERY RATING PLATE

The information indicated on the rating plate is the vehicle identification number, body type, exterior color, interior trim, date manufactured, D.S.O., rear axle and transmission type.

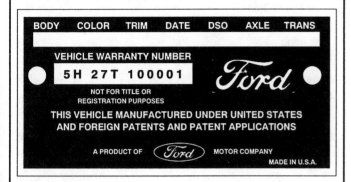

THE VEHICLE IDENTIFICATION NUMBER is a
series of letters and numbers on the rating plate. The VIN number identifies the model year, assembly plant, model type, engine and production sequence.

FIRST DIGIT: Identifies the model year (1965)

SECOND DIGIT: Identifies the assembly plant

ASSEMBLY PLANT	CODE
Atlanta, GA	A
Dallas, TX	D
Mahwah, NJ	E
Dearborn, MI	F
Chicago, IL	G
Lorain, OH	H
Los Angeles, CA	J
Kansas City, KS	K
Michigan Truck	L
Norfolk, VA	N
Twin Cities, MN	P
San Jose, CA	R
Allen Park, MI (Pilot Plant)	S
Metuchen, NJ	T
Louisville, KY	U
Wayne, MI	W
Wixom, MI	Y
St. Louis, MO	Z

THIRD AND FOURTH DIGITS: Identifies the model

MODEL	CODE
Ranchero	27
Sedan delivery	29

FIFTH DIGIT: Identifies the engine

ENGINE	CODE
289 cid, 8 cyl. (premium)	A
289 cid, 8 cyl.	C
200 cid, 6 cyl.	T
170 cid, 6 cyl.	U
200 cid, 6 cyl.	2
289 cid, 8 cyl.	3
170 cid, 6 cyl.	4

LAST SIX DIGITS: Identify the consecutive unit number

THE VEHICLE DATA appears on the line preceding the vehicle identification number.

THE BODY CODE indicates the body type.

BODY	CODE
2-Door standard	66A
2-Door deluxe	66B
2-Door standard (RPO w/bucket)	66G
2-Door standard (RPO w/bucket)	66H
2-Door standard sedan delivery	78A
2-Door deluxe sedan delivery	78B

THE EXTERIOR COLOR CODE indicates the paint
color of the vehicle. A single letter code designates a solid body color and two letters denote a two-tone; the first letter, the lower color and the second letter, the upper color.

COLOR	CODE
Black	A
Med. Ivy Gold Metallic	C
Med. Turquoise Metallic	D
Dk. Blue Metallic	H
Lt. Beige Metallic	I
Red	J
Med. Gray Metallic	K
White	M
Lt. Peacock	O
Palomino Metallic	P
Dk. Ivy Green Metallic	R
Yellow	V
Maroon Metallic	X
Med. Blue Metallic	Y
Dk. Turquoise Metallic	5

THE INTERIOR TRIM CODE indicates the trim color and material used on the vehicle.

COLOR	VINYL	CLOTH	LEATHER	CODE
Beige	•			04
Blue	•			92
Lt. Blue	•			62,72
Med./Lt. Blue	•			32,52
Med./Lt. Blue	•	•		12,22,82
Red	•			35,55,65
Red	•			75,85,95
Red	•	•		15,25
Black	•			26,36,56
Black	•			66,76,86
Black	•			96
Black	•	•		16
Turquoise	•			67
Lt. Turquoise Met.	•			77,87
Lt./Turquoise Met.	•			82
Med./Lt. Turquoise	•			37
Med./Lt. Turquoise	•	•		27
Med. Turquoise	•	•		17
Palomino	•			39,59,99
Palomino	•	•		29
Med. Palomino	•			49,69,79
Med. Palomino	•			89
Med./Lt. Palomino	•	•		19
Ivy Gold	•			38,68,78
Lt. Ivy Gold Met.	•			88
Med./Lt. Ivy Gold	•	•		28
White/Blue	•			42
White/Red	•			45
White/Black	•			46
White/Ivy Gold	•			48
White Pearl/Red	•	•		F2
White Pearl/Black	•	•		F5
White Pearl/ Turquoise	•	•		F6
White Pearl/Gold	•	•		F8
White Pearl/ Palomino	•	•		F9

COLOR	VINYL	CLOTH	LEATHER	CODE
White Pearl/Blue	•			G2
White Pearl/Red	•			G5
White Pearl/Black	•			G6
White Pearl/ Turquoise	•			G7
White Pearl/Gold	•			G8
White Pearl/ Palomino	•			G9
White/Blue	•			H2
White/Red	•			H5
White/Black	•			H6
White/Turquoise	•			H7
White/Gold	•			H8
White/Palomino	•			H9

THE TRANSMISSION CODE indicates the transmission type installed in the vehicle.

TYPE	CODE
3-Speed manual	1
4-Speed manual	5
Dual range automatic (C-4)	6

THE REAR AXLE CODE indicates the ratio of the rear axle installed in the vehicle. A number designates a conventional axle, while a letter designates an Equa-Lock differential.

RATIO	CODE
2.83:1	2
3.20:1	3
3.50:1	5
2.80:1	6

THE DATE CODE indicates the date the vehicle was manufactured. A number signifying the date precedes the month code letter. A second year code letter will be used if the model exceeds 12 months.

MONTH	FIRST YEAR	SECOND YEAR
January	A	N
February	B	P
March	C	Q
April	D	R
May	E	S
June	F	T
July	G	U
August	H	V
September	J	W
October	K	X
November	L	Y
December	M	Z

THE D.S.O. CODE: Units built on a Domestic Special Order, Foreign Special Order, or other special orders will have the complete order number in this space. Also to appear in this space is the two-digit code number of the District which ordered the unit. If the unit is a regular production unit, only the District code number will appear.

DISTRICT	CODE
Boston	11
Buffalo	12
New York	13
Pittsburgh	14
Newark	15
Atlanta	21
Charlotte	22
Philadelphia	23
Jacksonville	24
Richmond	25
Washington	26
Cincinnati	31
Cleveland	32
Detroit	33
Indianapolis	34
Lansing	35
Louisville	36
Chicago	41
Fargo	42
Rockford	43
Twin Cities	44
Davenport	45
Denver	51
Des Moines	52
Kansas City	53
Omaha	54
St. Louis	55
Dallas	61
Houston	62
Memphis	63
New Orleans	64
Oklahoma City	65
Los Angeles	71
San Jose	72
Salt Lake City	73
Seattle	74
Ford of Canada	81
Government	83
Home Office Reserve	84
American Red Cross	85
Transportation Services	89
Export	90's

ENGINE SPECIFICATIONS:

ENGINE CODE	NO. CYL.	CID	HORSE-POWER	COMP. RATIO	CARB
U	6	170	105	9.1:1	1 BC
T	6	200	120	9.2:1	1 BC
C	8	289	200	9.3:1	2 BC
A	8	289	225	10.0:1	4 BC

1966 BRONCO

1966 RANCHERO

1966 F-250 PICKUP

1966 BRONCO ROADSTER

F-SERIES
RATING PLATE

The information indicated on the rating plate is the vehicle identification number, wheelbase, exterior color, model type, body type, transmission type, rear axle, maximum gross vehicle weight (lbs.), certified net horsepower, r.p.m. and D.S.O. numbers.

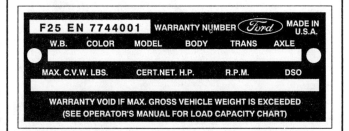

THE VEHICLE IDENTIFICATION NUMBER is a series of letters and numbers on the rating plate. The VIN number identifies the series, engine, assembly plant and production sequence.

FIRST, SECOND AND THIRD DIGITS: Identify the series

SERIES	CODE
F-100 (4x2)	F10
F-100 (4x4)	F11
F-250 (4x2)	F25
F-250 (4x4)	F26
F-350 (4x2)	F35

FOURTH DIGIT: Identifies the engine

ENGINE	CODE
240 cid, 6 cyl.	*A
300 cid, 6 cyl.	*B
352 cid, 6 cyl.	*Y
240 cid, 6 cyl.	**1
300 cid, 6 cyl.	**2
352 cid, 8 cyl.	**8

* Gas

** Low Compression

FIFTH DIGIT: Identifies the assembly plant

ASSEMBLY PLANT	CODE
Ontario, CAN	C
Dallas, TX	D
Mahway, NJ	E
Lorain, OH	H
Kansas City, KS	K
Michigan Truck	L
Norfolk, VA	N
Twin Cities, MN	P
San Jose, CA	R
Allen Park, MI (Pilot Plant)	S
Louisville, KY	U

LAST SIX DIGITS: Identifies the consecutive unit number

CALENDER YEAR 1965	NUMBERS
September	746,000 thru 759,999
October	760,000 thru 773,999
November	774,000 thru 787,999
December	788,000 thru 801,999
January	802,000 thru 815,999
February	816,000 thru 829,999
March	830,000 thru 843,999
April	844,000 thru 857,999
May	858,000 thru 871,999
June	872,000 thru 885,999
July	886,000 thru 899,999
August	900,000 thru 913,999

THE VEHICLE DATA appears on the two lines following the vehicle identification number.

THE W.B. (WHEELBASE) CODE indicates the wheelbase in inches.

THE EXTERIOR COLOR CODE indicates the paint color used on the vehicle.

COLOR	CODE
Raven Black	A
CaribbeanTurquoise	B
Pure White	C
Arcadian Blue	F
Chrome Yellow	G
Sahara Beige	H
Rangoon Red	J
Holly Green	L
Wimbledon White	M
Marlin Blue	W
Springtime Yellow	8

THE MODEL CODE indicates the model type and the gross vehicle weight (lbs.) information.

MODEL	GVW
F-100	5,000
F-101	4,200
F-102*	5,000
F-110 (4x4)	5,600
F-111 (4x4)	4,900
F-112* (4x4)	5,600
F-250	7,500
F-251	4,800
F-252*	7,500
F-260 (4x4)	6,800
F-261 (4x4)	4,900
F-262 (4x4)	7,700
F-350	10,000
F-351	8,000

* Reference Pennsylvania registration data

THE BODY CODE indicates the interior trim color and material, the second and third digits indicate the body type.

BODY:

TYPE	CODE
Conventional cab	81
Cowl and chassis	84
Cowl and windshield	85

THE INTERIOR TRIM CODE indicates the key to the trim color and material used on the vehicle.

COLOR	VINYL	WOVEN PLASTIC	LEATHER	CODE
Gray	•			1,J
Gray	•	•		A
Blue	•			2
Blue	•	•		B,K
Green	•			3
Green	•	•		C,L
Beige	•			4
Beige	•	•		D,M
Red	•			5,V
Red	•	•		E,N
Black	•			6,O,T
Parchment	•			U

THE TRANSMISSION CODE indicates the transmission type installed in the vehicle.

TYPE	CODE
4-Speed New Process	A
3-Speed Overdrive	B
3-Speed Ford (LD)	C
3-Speed Warner (MD)	D
3-Speed Warner (HD)	E
4-Speed Syn. Warner	F
Automatic (C-4)	G

THE REAR AXLE CODE indicates the ratio of the rear axle installed in the vehicle.

RATIO	CODE
4.11 - Ford	05
3.50 - Ford	08
3.70 - Ford	09
3.25 - Ford	17
4.83 - Dana #70	22
5.13 - Dana #70	23
4.10 - Dana #60	24
4.56 - Dana #60	25
4.83 - Dana #60	26
5.87 - Dana #70	29
3.54 - Dana #44	A8
3.54 - Dana #60-2	A9
4.10 - Dana #60	B4
4.56 - Dana #60	B5
4.88 - Dana #60	B6
3.31 - Dana #44	C1
3.73 - Dana #44	C2
4.09 - Dana #44	C4
4.10 - Dana #60-2	C5

FRONT AXLE CODES

CAPACITY	CODE
5.5M	A
5.5M	B
6M	C
7M	D
9M	E
11M	F
12M	G
15M	H
18M	I
6M	*L
6M, 7M	*M
9M	*N

* Heavy duty front brakes

THE MAX. G.V.W. LBS. CODE indicates the maximum gross vehicle weight in pounds.

THE CERT. NET. H.P. AT R.P.M. CODE indicates the certified net horsepower at specified r.p.m..

THE D.S.O. CODE: If the vehicle is built on a Direct Special Order the complete order number will be reflected under the DSO space, including the District Code Number.

DISTRICT	CODE
Boston	11
Buffalo	12
New York	13
Pittsburgh	14
Newark	15
Atlanta	21
Charlotte	22
Philadelphia	23
Jacksonville	24
Richmond	25
Washington	26
Cincinnati	31
Cleveland	32
Detroit	33
Indianapolis	34
Lansing	35
Louisville	36
Chicago	41
Milwaukee	42
Twin Cities	44
Davenport	45
Denver	51
Des Moines	52
Kansas City	53
Omaha	54
St. Louis	55
Dallas	61
Houston	62
Memphis	63
New Orleans	64
Oklahoma City	65
Los Angeles	71
San Jose	72
Salt Lake City	73
Seattle	74
Phoenix	75
Ford of Canada	81
Government	83
Home Office Reserve	84
American Red Cross	85
Transportation	89
Export	90's

ENGINE SPECIFICATIONS

ENGINE CODE	NO. CYL.	CID	HORSE-POWER	COMP. RATIO	CARB
A	6	240	150	9.2:1	1 BC
B	6	300	170	8.0:1	1 BC
8	8	352	208	8.9:1	2 BC

BRONCO RATING PLATE

The information indicated on the rating plate is the vehicle identification number, wheelbase, exterior color, model type, body type, transmission type, rear axle, maximum gross vehicle weight (lbs.), certified net horsepower, r.p.m. and D.S.O. numbers.

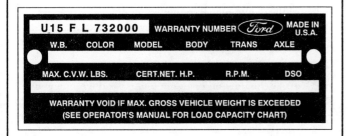

THE VEHICLE IDENTIFICATION NUMBER is a series of letters and numbers on the rating plate. The VIN number identifies the series, engine, assembly plant and production sequence.

FIRST, SECOND AND THIRD DIGITS: Identify the series

SERIES	CODE
U-100	U13
U-100	U14
U-100	U15

FOURTH DIGIT: Identifies the engine

ENGINE	CODE
240 cid, 6 cyl.	A
170 cid, 6 cyl.	F

FIFTH DIGIT: Identifies the assembly plant

ASSEMBLY PLANT	CODE
Lorain, OH	H
Michicgan Truck	L
San Jose, CA	R
Allen Park, MI (Pilot Plant)	S

LAST SIX DIGITS: Identify the consecutive unit number

CALENDAR YEAR	NUMBERS
August	732,000 thru 745,999
September	746,000 thru 759,999
October	760,000 thru 773,999
November	774,000 thru 787,999
December	788,000 thru 801,999
January	802,000 thru 815,999
February	816,000 thru 829,999
March	830,000 thru 843,999
April	844,000 thru 857,999
May	858,000 thru 871,999
June	872,000 thru 885,999
July	886,000 thru 899,999
August	900,000 thru 913,999
August	A00,000 thru A13,999

THE VEHICLE DATA appears on the two lines following the vehicle identification number.

THE W.B. (WHEELBASE) CODE indicates the wheelbase in inches.

THE EXTERIOR COLOR CODE indicates the paint color used on the vehicle.

COLOR	CODE
Raven Black	A
Caribbean Turquoise	B
Arcadian Blue	F
Sahara Beige	H
Rangoon Red	J
Holly Green	L
Wimbledon White	M
Marlin Blue	W
Springtime Yellow	8

THE MODEL CODE indicates the model type.

TYPE	CODE
U-100 (open body)	U13
U-100 (pickup)	U14
U-100 (wagon)	U15

THE BODY CODE indicates the body type and the interior trim color and material. The letter identifies the interior trim scheme and the two numerals identify the body or cab type.

TYPE	CODE
Open body (Roadster)	96
Pickup (short roof)	97
Wagon (long roof)	98

THE INTERIOR TRIM CODE indicates the key to the trim color and material used on the vehicle.

INTERIOR TRIM	BENCH	BUCKET DRIVER ONLY	BUCKET DRIVER & PASS.
Gray	1	6	7

THE TRANSMISSION CODE indicates the transmission type installed in the vehicle.

TYPE	CODE
3-Speed manual shift	C

THE REAR AXLE CODE indicates the ratio of the rear axle installed in the vehicle. A number "2" immediately following a rear axle code on the rating plate indicates vehicle equipped with a locking front axle.

RATIO	REGULAR	LOCKING
4.11:1	03	—
4.57:1	04	—
4.11:1	05	A5
4.57:1	06	A6

THE D.S.O. CODE: If the vehicle is built on a Direct Special Order, Foreign Special Order, or other special orders, the complete order number will be reflected under the D.S.O. space after the District Code Number.

DISTRICT	CODE
Boston	11
New York	13
Newark	15
Philadelphia	16
Washington	17
Atlanta	21
Charlotte	22
Jacksonville	24
Richmond	25
Cincinnati	27
Louisville	28
Cleveland	32
Detroit	33
Indianapolis	34
Lansing	35
Buffalo	37
Pittsburgh	38
Chicago	41
Fargo	42
Milwaukee	43
Twin Cities	44
Davenport	45
Denver	51
Des Moines	52
Kansas City	53
Omaha	54
St. Louis	55
Dallas	61
Houston	62
Memphis	63
New Orleans	64
Oklahoma City	65
Los Angeles	71
San Jose	72
Salt Lake City	73
Seattle	74
Phoenix	75
Ford of Canada	81
Government	83
Home Office Reserve	84
American Red Cross	85
Transportation Services	89
Export	90's

ENGINE SPECIFICATIONS

ENGINE CODE	NO. CYL.	CID	HORSE-POWER	COMP. RATIO	CARB
F	6	170	105	9.1:1	1 BC

RANCHERO
RATING PLATE

The information indicated on the rating plate is the vehicle identification number, body type, exterior color, interior trim, date manufactured, D.S.O., rear axle and transmission type.

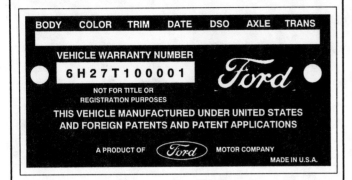

THE VEHICLE IDENTIFICATION NUMBER is a
series of letters and numbers on the rating plate. The VIN number identifies the model year, assembly plant, model type, engine and production sequence.

FIRST DIGIT: Identifies the model year (1966)

SECOND DIGIT: Identifies the assembly plant

ASSEMBLY PLANT	CODE
Atlanta, GA	A
Oakville, CAN	B
Ontario Truck	C
Dallas, TX	D
Mahwah, NJ	E
Dearborn, MI	F
Chicago, IL	G
Lorain, OH	H
Los Angeles, CA	J
Kansas City, KS	K
Michigan Truck	L
Norfolk, VA	N
Twin Cities, MN	P
San Jose, CA	R
Allen Park, MI (Pilot Plant)	S
Metuchen, NJ	T
Louisville, KY	U
Wayne, MI	W
Wixom, MI	Y
St. Louis, MO	Z

THIRD AND FOURTH DIGITS: Identify the model

MODEL	CODE
Ranchero	27

FIFTH DIGIT: Identifies the engine

ENGINE	CODE
289 cid, 8 cyl.	A
289 cid, 8 cyl.	C
200 cid, 6 cyl.	T

LAST SIX DIGITS: Identify the consecutive unit number

THE VEHICLE DATA appears on the line preceding the vehicle identification number.

THE BODY CODE indicates the body type.

TYPE	CODE
2-Door standard	66A
2-Door deluxe	66B
2-Door standard (RPO bucket w/console)	66D

THE EXTERIOR COLOR CODE indicates the paint color used on the vehicle.
A single letter code designates a solid body color and two letters denote a two-tone; the first letter, the lower color and the second letter, the upper color.

COLOR	CODE
Raven Black	A
Arcadian Blue	F
Sahara Beige	H
Nightmist Blue	K
Wimbledon White	M
Antique Bronze	P
Ivy Green	R
Candyapple Red	T
Tahoe Turquoise	U
Emberglo Metallic	V
Vintage Burgundy	X
Silver Blue	Y
Sauterne Gold Metallic	Z
Frost Silver	4
Springtime Yellow	8

THE INTERIOR TRIM CODE indicates the key to the interior trim color and material used on the vehicle.

COLOR	RANCHERO	CUSTOM BENCH	RANCHERO BUCKET
Blue vinyl	—	42	82
Red vinyl	—	45	85
Black vinyl	26	46	86
Parchment vinyl	2D	4D	8D

THE DATE CODE indicates the date the vehicle was manufactured. A number signifying the date precedes the month code letter. A second-year code letter will be used if the model exceeds 12 months.

MONTH	FIRST YEAR	SECOND YEAR
January	A	N
February	B	P
March	C	Q
April	D	R
May	E	S
June	F	T
July	G	U
August	H	V
September	J	W
October	K	X
November	L	Y
December	M	Z

THE D.S.O. CODE: Units built on a Domestic Special Order, Foreign Special Order, or other special orders will have the complete order number in this space. Also to appear in this space is the two-digit code number of the District which ordered the unit. If the unit is a regular production unit, only the District code number will appear.

DISTRICT	CODE
Boston	11
Buffalo	12
New York	13
Pittsburgh	14
Newark	15
Atlanta	21
Charlotte	22
Philadelphia	23
Jacksonville	24
Richmond	25
Washington	26
Cincinnati	31

Cleveland	32
Detroit	33
Indianapolis	34
Lansing	35
Louisville	36
Chicago	41
Fargo	42
Rockford	43
Twin Cities	44
Davenport	45
Denver	51
Des Moines	52
Kansas City	53
Omaha	54
St. Louis	55
Dallas	61
Houston	62
Memphis	63
New Orleans	64
Oklahoma City	65
Los Angeles	71
San Jose	72
Salt Lake City	73
Seattle	74
Ford of Canada	81
Government	83
Home Office Reserve	84
American Red Cross	85
Transportation Services	89
Export	90's

THE TRANSMISSION CODE indicates the transmission type installed in the vehicle.

TYPE	CODE
3-Speed manual	1
Dual range automatic (C-6)	4
4-Speed manual	5
Dual range automatic (C-4)	6

THE REAR AXLE CODE indicates the ratio of the rear axle installed in the vehicle.

RATIO	CODE
3.00	1
3.25	4
3.50	5
3.00	A
3.25	D
3.50	E

ENGINE SPECIFICATIONS:

ENGINE CODE	NO. CYL.	CID	HORSE-POWER	COMP. RATIO	CARB
T	6	200	120	9.2:1	1 BC
C	8	289	200	9.3:1	2 BC
A	8	289	225	10.0:1	4 BC

1967 F-100 RANGER

1967 F-100 RANGER

1967 BRONCO

1967 FAIRLANE RANCHERO

1967 F-250 SUPERCAB

1967 F-250 PICKUP

F-SERIES
RATING PLATE

The information indicated on the rating plate is the vehicle identification number, wheelbase, exterior color, model type, body type, transmission type, rear axle, maximum gross vehicle weight in lbs., certified net horsepower, r.p.m. and D.S.O. numbers.

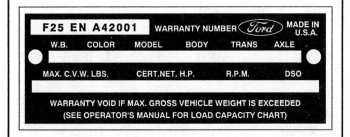

THE VEHICLE IDENTIFICATION NUMBER is a series of letters and numbers on the rating plate. The VIN number identifies the series, engine, assembly plant and production sequence.

FIRST, SECOND AND THIRD DIGITS: Identify the series

SERIES	CODE
F-100 (4x2)	F10
F-100 (4x4)	F11
F-250 (4x2)	F25
F-250 (4x4)	F26
F-350 (4x2)	F35

FOURTH DIGITS: Identifies the engine

ENGINE	CODE
240 cid, 6 cyl.	A
300 cid, 6 cyl. (LD)	B
300 cid, 6 cyl. (HD)	B
352 cid, 8 cyl.	Y
300 cid, 6 cyl.	*2
352 cid, 8 cyl.	*8

* Low compression

FIFTH DIGIT: Identifies the assembly plant

ASSEMBLY PLANT	CODE
Ontario, CAN	C
Dallas, TX	D
Mawah, NJ	E
Lorain, OH	H
Kansas City	K
Michigan Truck	L
Norfolk, VA	N
Twin Cities, MN	P
San Jose, CA	R
Allen Park, MI (Pilot Plant)	S
Louisville, KY	U

LAST SIX DIGITS: Identifies the consecutive unit number

CALENDAR YEAR 1966	NUMBERS
August - 1967 Model	A00,000 thru A13,999
September	A14,000 thru A27,999
October	A28,000 thru A41,999
November	A42,000 thru A55,999
December	A56,000 thru A69,999

CALENDAR YEAR 1967	
January	A70,000 thru A83,999
February	A84,000 thru A97,999
March	A98,000 thru B11,999
April	B12,000 thru B25,999
May	B26,000 thru B39,999
June	B40,000 thru B53,999
July	B54,000 thru B67,999
August	B68,000 thru B81,999

THE VEHICLE DATA appears on the two lines following the vehicle identification number.

THE W.B. (WHEELBASE) CODE indicates the wheelbase in inches.

THE EXTERIOR COLOR CODE indicates the paint color used on the vehicle.

COLOR	CODE
Raven Black	A
Frost Turquoise	B
Pure White	C
Chrome Yellow	G
Rangoon Red	J
Holly Green	L
Wimbledon White	M
Lunar Green	U
Pebble Beige	6
Harbor Blue	7
Springtime Yellow	8
Prime	9

THE MODEL CODE indicates the model type and the gross vehicle weight (lbs.) information.

MODEL	GVW
F-100	5,000
F-101	4,200
F-102*	5,000
F-103	4,500
F-110 (4x4)	5,600
F-111 (4x4)	4,900
F-112* (4x4)	5,600
F-250	7,500
F-251	4,800
F-252*	7,500
F-253	6,000
F-254	6,900
F-255*	6,000
F-256*	6,900
F-260 (4x4)	6,800
F-261 (4x4)	4,900
F-262 (4x4)	7,700
F-263 (4x4)	6,100
F-264* (4x4)	7,700
F-350	10,000
F-351	8,000
F-352*	8,000
F-353	6,600

* Reference Pennsylvania registration data

THE BODY CODE: The first digit identifies the interior trim and the second and third digits indicate the body style.

BODY STYLE	CODE
Conventional cab	81
Cowl and chassis	84
Cowl and windshield	85

INTERIOR TRIM CODES:

COLOR	VINYL	WOVEN PLASTIC	LEATHER	CODE
Med. Blue/Dk. Blue	•			F
Dk. Blue	•			O,S,W
Med. Green/Dk. Green	•			G
Dk. Green	•			P,T,X
Red	•			V
Red/Dk. Red	•			I
Dk. Red	•			R,Z
Lt. Parchment	•			H,R,U
Lt. Parchment	•			Y

THE TRANSMISSION CODE indicates the transmission type installed in the vehicle.

TYPE	CODE
4-Speed new process	A
3-Speed overdrive	B
3-Speed Ford L.D.	C
3-Speed Warner M.D.	D
3-Speed Warner H.D.	E
4-Speed Syn. Warner	F
Automatic (C-4)	G

THE REAR AXLE CODE indicates the ratio of the rear axle installed in the vehicle.

RATIO	CODE
4.11 - Ford	05
3.50 - Ford	08
3.70 - Ford	09
3.25 - Ford	17
4.88 - Dana #70	22
4.10 - Dana #60	24
4.56 - Dana #60	25
4.88 - Dana #60	26
4.10 - Dana #70	27

RATIO	CODE
4.56 - Dana #70	28
3.73 - Dana #60	38
3.54 - Dana #44	A8
3.54 - Dana #60-2	A9
4.10 - Dana #60	B4
4.56 - Dana #60	B5
4.88 - Dana #60	B6
3.31 - Dana #44	C1
3.73 - Dana #44	C2
4.09 - Dana #44	C4
4.10 - Dana #60-2	C5
3.73 - Dana #60	C8
4.88 - Dana #70	*D2
4.10 - Dana #70	*D7
4.56 - Dana #70	*D8
4.10 - Dana #60-3	*E5
3.54 - Dana #60-3	*E9

* Locking

FRONT AXLE:

CAPACITY	CODE
5.5M	B
6M	C
7M	D
9M	E
12M	F
12M	G
15M	H
18M	I
3.5M	K
6M	*L
Heavy duty front brakes	M
9M	*N

* Heavy duty front brakes

THE MAX. G.V.W. LBS. CODE indicates the maximum gross vehicle weight in pounds.

THE CERT. NET. H.P. CODE indicates the certified net horsepower at specified rpm.

THE D.S.O. CODE: If the vehicle is built on a Direct Special Order, the complete order number will be reflected under the DSO space including the District Code Number.

DISTRICT	CODE
Boston	11
New York	13
Newark	15
Philadelphia	16
Washington	17
Atlanta	21
Charlotte	22
Jacksonville	24
Richmond	25
Cincinnati	27
Louisville	28
Cleveland	32
Detroit	33
Indianapolis	34
Lansing	35
Buffalo	37
Pittsburgh	38
Chicago	41
Fargo	42
Milwaukee	43
Twin Cities	44
Davenport	45
Denver	51
Des Moines	52
Kansas City	53
Omaha	54
St. Louis	55
Dallas	61
Houston	62
Memphis	63
New Orleans	64
Oklahoma City	65
Los Angeles	71
San Jose	72
Salt Lake City	73
Seattle	74
Phoenix	75
Ford of Canada	81
Government	83
Home Office Reserve	84
American Red Cross	85
Transportation	89
Export	90's

ENGINE SPECIFICATIONS:

ENGINE CODE	NO. CYL.	CID	HORSE-POWER	COMP. RATIO	CARB
A	6	240	150	9.2:1	1 BC
B	6	300	170	8.0:1	1 BC
Y	8	352	208	8.9:1	2 BC

BRONCO RATING PLATE

The information indicated on the rating plate is the vehicle identification number, the wheelbase, the exterior color, the model type, the body type, the transmission type, the rear axle, the maximum gross vehicle weight (lbs.), the certified net horsepower, the r.p.m. and the D.S.O. numbers.

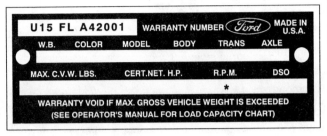

* Not used on the Bronco

THE VEHICLE IDENTIFICATION NUMBER is a series of letters and numbers on the rating plate. The VIN number identifies the series, engine, assembly plant and production sequence.

FIRST, SECOND AND THIRD DIGITS: Identify the series

SERIES	CODE
U-100 (roadster)	U13
U-100 (pickup)	U14
U-100 (wagon)	U15

FOURTH DIGIT: Identifies the engine

	LOW COMPRESSION	
ENGINE	CODE	CODE
170 cid, 6 cyl.	6	F
289 cid, 8 cyl.	-	N

FIFTH DIGIT: Identifies the assembly plant

ASSEMBLY PLANT	CODE
Lorain, OH	H
Michigan Truck	L
San Jose, CA	R
Allen Park, MI (Pilot Plant)	S

LAST SIX DIGITS: Identify the consecutive unit number

MONTH	NUMBERS
August 1967 model	A00,000 - A13,999
September	A14,000 - A27,999
October	A28,000 - A41,999
November	A42,000 - A55,999
December	A56,000 - A69,999
January	A70,000 - A83,999
February	A84,000 - A97,999
March	A98,000 - A111,999
April	A112,000 - A125,999
May	A126,000 - A139,999
June	A140,000 - A153,999
July	A154,000 - A167,999
August	A168,000 - A171,999

THE VEHICLE DATA appears on the two lines following the vehicle identification number.

THE W.B. (WHEELBASE) CODE indicates the wheelbase in inches.

THE EXTERIOR COLOR CODE indicates the paint color used on the vehicle.

COLOR	CODE
Raven Black	A
Frost Turquoise	B
Peacock Blue	D
Chrome Yellow	G
Rangoon Red	J
Holly Green	L
Wimbledon White	M
Poppy Red	S
Lunar Green	U
Pebble Beige	6
Harbor Blue	7
Springtime Yellow	8
Prime	9

THE MODEL CODE indicates the model type.

MODEL U-100	CODE
U13 Open body (roadster)	96
U14 Pickup (short roof)	97
U15 Wagon (long roof)	98

THE BODY CODE indicates the key to the trim color and material.
INTERIOR TRIM:

COLOR	VINYL	WOVEN PLASTIC	LEATHER	CODE
Parchment	•			1,7,8,9

THE TRANSMISSION CODE indicates the transmission type installed in the vehicle.

TYPE	CODE
3-Speed manual shift	C
C-4 Automatic	G

THE REAR AXLE CODE indicates the ratio of the rear axle installed in the vehicle.

RATIO	REGULAR	LOCKING
4.11:1	03	A3
4.57:1	04	A4
4.11:1	05	A5
4.57:1	06	A6
3.50:1	18	B8
3.50:1	19	B9

Note: The number 2 immediately following a rear axle code on the rating plate indicates vehicle equipped with a locking front axle.

THE D.S.O. CODE: If the vehicle is built on a Direct Special Order the complete order number will be reflected under the DSO space after the District Code Number.

DISTRICT	CODE
Boston	11
New York	13
Newark	15
Philadelphia	16
Washington	17
Atlanta	21
Charlotte	22
Jacksonville	24
Richmond	25
Cincinnati	27
Louisville	28
Cleveland	32
Detroit	33
Indianapolis	34
Lansing	35
Buffalo	37
Pittsburgh	38
Chicago	41
Fargo	42
Milwaukee	43
Twin Cities	44
Davenport	45
Denver	51
Des Moines	52
Kansas City	53
Omaha	54
St. Louis	55
Dallas	61
Houston	62
Memphis	63
New Orleans	64
Oklahoma City	65
Los Angeles	71
San Jose	72
Salt Lake City	73
Seattle	74
Phoenix	75
Ford of Canada	81
Government	83
Home Office Reserve	84
American Red Cross	85
Transportation Services	89
Export	90's

ENGINE SPECIFICATIONS:

ENGINE CODE	NO. CYL.	CID	HORSE-POWER	COMP. RATIO	CARB
F	6	170	105	9.1:1	1 BC
N	8	289	200	9.3:1	2 BC

RANCHERO
RATING PLATE

The information indicated on the rating plate is the vehicle identification number, the body type, the exterior color, the interior trim, the date manufactured, the D.S.O., the rear axle and the transmission type.

7H4 7C 100001	WARRANTY NUMBER	Ford	MADE IN U.S.A

NOT FOR TITLE OR REGISTRATION

| BODY | COLOR | TRIM | DATE | DSO | AXLE | TRANS |

THE VEHICLE IDENTIFICATION NUMBER is a series of letters and numbers on the rating plate. The VIN number identifies the model year, assembly plant, series, engine and production sequence.

FIRST DIGIT: Identifies the model year (1967)

SECOND DIGIT: Identifies the assembly plant

ASSEMBLY PLANT	CODE
Atlanta, GA	A
Oakville, CAN	B
Ontario Truck	C
Dallas, TX	D
Mahwah, NJ	E
Dearborn, MI	F
Chicago, IL	G
Lorain, OH	H
Los Angeles, CA	J
Kansas City, KS	K
Michigan Truck	L
Norfolk, VA	N
Twin Cities, MN	P
San Jose, CA	R
Allen Park, MI (Pilot Plant)	S
Metuchen, NJ	T
Louisville, KY	U
Wayne, MI	W
Wixom, MI	Y
St. Louis, MO	Z

THIRD AND FOURTH DIGITS: Identify the body serial code

BODY	CODE
2-Door w/bench seat	47
2-Door w/bench seat	48
2-Door w/bucket seat	49

FIFTH DIGIT: Identifies the engine

ENGINE	CODE
200 cid, 6 cyl.	T
200 cid, 6 cyl.	*2
289 cid, 8 cyl.	C
289 cid, 8 cyl.	*3
289 cid, 8 cyl. (premium fuel)	A
289 cid, 8 cyl. (hi-perf.)	K
390 cid, 8 cyl.	Y
390 cid, 8 cyl.	H
390 cid, 8 cyl.	S

LAST SIX DIGITS: Identify the consecutive unit number

THE VEHICLE DATA appears on the line following the vehicle identification number.

THE BODY CODE indicates the body type.

TYPE	CODE
2-Door w/bench seat	66A
2-Door w/bench seat	66B
2-Door w/bucket seat	66D

THE EXTERIOR COLOR CODE indicates the paint color used on the vehicle.

COLOR	CODE
Raven Black	A
Frost Turquoise	B
Beige Mist	E
Nightmist Blue	K
Wimbledon White	M
Brittany Blue	Q
Candyapple Red	T
Burnt Amber	V
Clearwater Aqua	W
Vintage Burgandy	X
Dk. Moss Green	Y
Sauterne Gold	Z
Silver Frost	4
Pebble Beige	6
Springtime Yellow	8

THE INTERIOR TRIM CODE indicates the key to the trim color and material.

COLOR	VINYL	CLOTH	LEATHER	CODE
Black	•			2A,4A,8A
Blue	•			2B,4B,8B
Blue	•	•		5B
Red	◦			2D,4D,8D
Parchment	◦			4U,6U
Parchment/Black	•			2U,8U,FA

THE DATE CODE indicates the date the vehicle was manufactured. A number signifying the date precedes the month code letter. A second year code letter will be used if the model exceeds 12 months.

MONTH	FIRST YEAR	SECOND YEAR
January	A	N
February	B	P
March	C	Q
April	D	R
May	E	S
June	F	T
July	G	U
August	H	V
September	J	W
October	K	X
November	L	Y
December	M	Z

THE TRANSMISSION CODE indicates the transmission type installed in the vehicle.

TYPE	CODE
3-Speed manual	1
Overdrive	2
3-Speed manual	3
4-Speed manual	5
Automatic (C4)	W
Automatic (C6)	U

THE REAR AXLE CODE indicates the ratio of the rear axle installed in the vehicle. A number designates a conventional axle, while a letter designates a locking differential.

RATIO	CODE
3.00:1	1
2.83:1	2
3.20:1	3
3.25:1	4
3.50:1	5
2.80:1	6
3.00:1	A
3.20:1	C
3.25:1	D
3.50:1	E

THE D.S.O. CODE: Units built on a Domestic Special Order, Foreign Special Order, or other special orders will have the complete order number in this space. Also to appear in this space is the two-digit code number of the District which ordered the unit. If the unit is a regular production unit, only the District code number will appear.

DISTRICT	CODE
Boston	11
New York	13
Newark	15
Philadelphia	16
Washington	17
Atlanta	21
Charlotte	22
Jacksonville	24
Richmond	25
Cincinnati	27
Louisville	28
Cleveland	32
Detroit	33
Indianapolis	34
Lansing	35
Buffalo	37
Pittsburgh	38
Chicago	41
Fargo	42
Milwaukee	43
Twin Cities	44
Davenport	45
Denver	51
Des Moines	52
Kansas City	53
Omaha	54
St. Louis	55
Dallas	61
Houston	62
Memphis	63
New Orleans	64
Oklahoma City	65
Los Angeles	71
San Jose	72
Salt Lake City	73
Seattle	74
Phoenix	75
Ford of Canada	81
Government	83
Home Office Reserve	84
American Red Cross	85
Transportation Services	89
Export	90's

ENGINE SPECIFICATIONS:

ENGINE CODE	NO. CYL.	CID	HORSE-POWER	COMP. RATIO	CARB
T	6	200	120	9.2:1	1 BC
C	8	289	200	9.3:1	2 BC
A	8	289	225	10.0:1	4 BC
Y	8	390	265	9.5:1	2 BC
K	8	289	271	10.5:1	4 BC
H	8	390	275	9.5:1	2 BC
S	8	390	315	10.5:1	4 BC

1968 BRONCO

1968 RANCHERO 500

1968 FAIRLANE RANCHERO GT

1968 F-100 RANGER

1968 F-100 RANGER

1968 F-100 RANGER

F-SERIES
RATING PLATE

The information indicated on the rating plate is the vehicle identification number, the wheelbase, exterior color, model type, body type, transmission type, rear axle, maximum gross vehicle weight in lbs., certified net horsepower, r.p.m. and the D.S.O. numbers.

THE VEHICLE IDENTIFICATION NUMBER is a series of letters and numbers on the rating plate. The VIN number identifies the series, engine, assembly plant and production sequence.

FIRST, SECOND AND THIRD DIGITS: Identify the series

SERIES	CODE
F-100 (4x2)	F10
F-100 (4x4)	F11
F-250 (4x2)	F25
F-250 (4x4)	F26
F-350 (4x2)	F35

FOURTH DIGIT: Identifies the engine

ENGINE	CODE
240 cid, 6 cyl.	A
300 cid, 6 cyl.	B
390 cid, 8 cyl.	H
360 cid, 8 cyl.	Y

FIFTH DIGIT: Identifies the assembly plant

ASSEMBLY PLANT	CODE
Ontario, CAN	C
Dallas, TX	D
Mawah, NJ	E
Lorain, OH	H
Kansas City, KS	K
Michigan Truck	L
Norfolk, VA	N
Twin Cities, MN	P
San Jose, CA	R
Allen Park, MI (Pilot Plant)	S
Louisville, KY	U

LAST SIX DIGITS: Identify the consecutive unit number

CALENDER YEAR 1967	NUMBERS
August - 1968 model	C00,000 thru C13,999
September	C14,000 thru C27,999
October	C28,000 thru C41,999
November	C42,000 thru C55,999
December	C56,000 thru C69,999

CALENDAR YEAR 1968	
January	C70,000 thru C83,999
February	C84,000 thru C97,999
March	C98,000 thru D11,999
April	D12,000 thru D25,999
May	D26,000 thru D39,999
June	D40,000 thru D53,999
July	D54,000 thru D67,999
August	D68,000 thru D81,999

THE VEHICLE DATA appears on the two lines following the vehicle identification number.

THE W.B. (WHEELBASE) CODE indicates the wheelbase in inches.

THE EXTERIOR COLOR CODE indicates the paint color used on the vehicle.

COLOR	CODE
Raven Black	A
Pure White	C
Sky View Blue	E
Chrome Yellow	G
Rangoon Red	J
Holly Green	L
Wimbledon White	M
Lunar Green	U
Meadowlark Yellow	W
Pebble Beige	6
Harbor Blue	7
Prime	9

THE MODEL CODE indicates the model type and the gross vehicle weight (lbs.) information.

MODEL	GVW
F-100	5,000
F-101	4,200
F-102*	5,000
F-103	4,500
F-104	4,800
F-110 (4x4)	5,600
F-111 (4x4)	5,000
F-112* (4x4)	5,600
F-113 (4x4)	4,600
F-250	7,500
F-252*	7,500
F-253	6,100
F-254	6,900
F-255*	6,100
F-256*	6,900
F-260 (4x4)	6,800
F-262 (4x4)	7,700
F-263 (4x4)	6,300
F-264* (4x4)	7,700
F-350	10,000
F-351	8,000
F-352*	8,000
F-353	6,600
F-354	8,300
F-355	9,000

* Reference Pennsylvania registration data

THE BODY CODE indicates the key to the interior trim color and material used in the vehicle.

COLOR	VINYL	WOVEN PLASTIC	LEATHER	CODE
Gray	•			1
Med. Blue	•			2
Med. Blue	•	•		23,B
Dk. Blue	•			24,B4,K4
Med. Beige	•			3
Med. Beige	•	•		33,C
Black	•			44,D4,M4
Black	•			4,M
Black	•	•		43,D
Red	•			54,E4,N4
Red	•			5
Red	•	•		53,E
Parchment	•			14,A4,J4
Parchment	•			J
Gray/Multicolor	•	•		A,13
Med./Dk. Blue	•			K
Red/Dk. Red	•			N

BODY STYLE	CODE
Conventional cab	81
Cowl and chassis	84
Cowl and windshield	85

THE TRANSMISSION CODE indicates the transmission type installed in the vehicle.

TYPE	CODE
4-Speed new process	A
3-Speed overdrive	B
3-Speed Ford L.D.	C
3-Speed Warner M.D.	D
3-Speed Warner H.D.	E
4-Speed Syn. Warner	F
Automatic (C-4)	G
4-Speed Warner	P

THE REAR AXLE CODE: The first two digits represent the rear axle, the third represents the front axle, if applicable.

RATIO & RATING	CODE
4.11 - Ford	05
3.50 - Ford	08
3.70 - Ford	09
3.25 - Ford	17
4.88 - Dana #70	22
5.13 - Dana #70	23
4.10 - Dana #60	24
4.56 - Dana #60	25
4.10 - Dana #70	27
4.56 - Dana #70	28
5.29 - Rockwell C-100	30
6.20 - Rockwell C-100	32
6.80 - Rockwell C-100	34
3.73 - Dana #70	36
3.54 - Dana #60	37
3.73 - Dana #60	38
5.83 - Rockwell D-100W	41
6.20 - Rockwell D-100N	42
6.80 - Rockwell D-100N	44
3.54 - Dana #44	*A8
4.10 - Dana #60	*B4
4.56 - Dana #60	*B5
3.31 - Dana #44-3	*C1
3.73 - Dana #44-3	*C2
4.09 - Dana #44-3	*C4
4.10 - Dana #60-2	*C5
3.54 - Dana #60	*C7
3.73 - Dana #60	*C8
4.88 - Dana #70	*D2
3.73 - Dana #70	*D6
4.10 - Dana #70	*D7
4.56 - Dana #70	*D8
4.10 - Dana #60-3	*E5
3.54 - Dana #60-3	*E9

* Locking

FRONT AXLE:

CAPACITY	CODE
5.5M	B
6M	C
7M	D
9M	E
12M	F
12M	G
15M	H
18M	I
3.5M	K
6M	*L

* Heavy duty front brakes

THE MAX. G.V.W. LBS. CODE indicates the maximum gross vehicle weight in pounds.

THE CERT. NET. H.P. R.P.M. indicates the certified net horsepower at specified rpm.

THE D.S.O. CODE: If the vehicle is built on a Direct Special Order, the complete order number will be reflected under the DSO space, including the District Code Number.

DISTRICT	CODE
Boston	11
New York	13
Newark	15
Philadelphia	16
Washington	17
Atlanta	21
Charlotte	22
Jacksonville	24
Richmond	25
Cincinnati	27
Louisville	28
Cleveland	32
Detroit	33
Indianapolis	34
Lansing	35
Buffalo	37
Pittsburgh	38
Chicago	41
Fargo	42
Milwaukee	43
Twin Cities	44
Davenport	45
Denver	51
Des Moines	52
Kansas City	53
Omaha	54
St. Louis	55
Dallas	61
Houston	62
Memphis	63
New Orleans	64
Okalahoma City	65
Los Angeles	71
San Jose	72
Salt Lake City	73
Seattle	74
Phoenix	75
Government	83
Home Office Reserve	84
American Red Cross	85
Transportation	89
Export	90's

FORD OF CANADA

Central	B1
Eastern	B2
Atlantic	B3
Midwestern	B4
Western	B6
Pacific	B7
Export	I1 thru I7

ENGINE SPECIFICATIONS

ENGINE CODE	NO. CYL.	CID	HORSE-POWER	COMP. RATIO	CARB
A	6	240	150	9.2:1	1 BC
B	6	300	165	8.8:1	1 BC
Y	8	360	215	8.4:1	2 BC
H	8	390	255	8.6:1	2 BC

BRONCO
RATING PLATE

The information indicated on the rating plate is the vehicle identification number, wheelbase, exterior color, model type, body type, transmission type, rear axle, maximum gross vehicle weight (lbs.), certified net horsepower, r.p.m. and D.S.O. numbers.

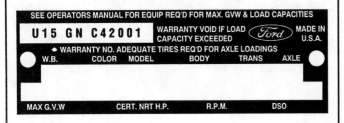

THE VEHICLE IDENTIFICATION NUMBER is a series of letters and numbers on the rating plate. The VIN number identifies the series, engine, assembly plant and production sequence.

FIRST, SECOND AND THIRD DIGITS: Identify the series

SERIES	CODE
U-130 (roadster)	U13
U-140 (sports utility)	U14
U-150 (wagon)	U15

FOURTH DIGIT: Identifies the engine

ENGINE	LOW COMPRESSION CODE	CODE
170 cid, 6 cyl.	6	F
240 cid, 6 cyl.	1	A
289 cid, 8 cyl.	—	N
302 cid, 8 cyl.	7	G

FIFTH DIGIT: Identifies the assembly plant

ASSEMBLY PLANT	CODE
Lorain, OH	H
Michigan Truck	L
San Jose, CA	R
Allen Park, MI (Pilot Plant)	S

SIX DIGITS: Identify the consecutive unit number

CALENDAR YEAR	NUMBERS
November - 1967 model	C42,000 -C55,999
December	C56,000 -C69,999
January	C70,000 -C83,999
February	C84,000 -C97,999
March	C98,000 -C111,999
April	D12,000 -D25,999
May	D26,000 -D39,999
June	D40,000 -D53,999
July	D54,000 -D67,999
August	D68,000 -D81,999

THE VEHICLE DATA appears on the two lines following the vehicle identification number.

THE W.B. (WHEELBASE) CODE indicates the wheelbase in inches.

THE EXTERIOR COLOR CODE indicates the paint color used on the vehicle.

COLOR	CODE
Raven Black	A
Peacock Blue	D
Sky View Blue	E
Chrome Yellow	G
Rangoon Red	J
Holly Green	L
Wimbledon White	M
Lunar Green	U
Meadowlark Yellow	W
Signal Flare Red	Z
Pebble Beige	6
Harbor Blue	7

THE MODEL CODE indicates the model type.

MODEL	CODE
Open body (roadster)	U-130
HD package	U-132
Sports Utility	U-140
HD package	U-142
Long roof (wagon)	U-150
HD package	U-152

THE BODY CODE indicates the body type and the interior trim used on the vehicle.

BODY	CODE
Open body (roadster)	96
Sports utility	97
Long roof (wagon)	98

INTERIOR TRIM:

TYPE/SEATS	CODE
Parchment, w/vinyl bench	*1
Parchment, w/vinyl R/L bucket	94
Parchment, w/vinyl R/L bucket/bench	94
Parchment, w/vinyl sport bench	9

* Standard trim

THE TRANSMISSION CODE indicates the transmission type installed in the vehicle.

TYPE	CODE
3-Speed manual shift	C
C-4 automatic	G

THE REAR AXLE CODE indicates the ratio of the rear axle installed in the vehicle. The number "2" immediately following a rear axle code on the rating plate indicates a vehicle equipped with a locking front axle.

RATIO	REGULAR	LOCKING
4.11	03	A3
4.57	04	A4
4.11	05	A5
4.57	06	A6
3.50	18	B8
3.50	19	B9

THE D.S.O. CODE: If the vehicle is built on a Direct Special Order, the complete order number will be reflected over the DSO space after the District Code Number.

DISTRICT	CODE
Boston	11
New York	13
Newark	15
Philadelphia	16
Washington	17
Atlanta	21
Charlotte	22
Jacksonville	24
Richmond	25
Cincinnati	27
Louisville	28
Cleveland	32
Detroit	33
Indianapolis	34
Lansing	35
Buffalo	37
Pittsburgh	38
Chicago	41
Fargo	42
Milwaukee	43
Twin Cities	44
Davenport	45
Denver	51
Des Moines	52
Kansas City	53
Omaha	54
St. Louis	55
Dallas	61
Houston	62
Memphis	63
New Orleans	64
Oklahoma City	65
Los Angeles	71
San Jose	72
Salt Lake City	73
Seattle	74
Phoenix	75
Ford of Canada	81
Government	83
Home Office Reserve	84
American Red Cross	85
Transportation Services	89
Export	90's

ENGINE SPECIFICATIONS

ENGINE CODE	NO. CYL.	CID	HORSE-POWER	COMP. RATIO	CARB
F	6	170	100	8.7:1	1 BC
A	6	240	150	9.2:1	1 BC
N	8	289	200	9.3:1	2 BC
G	8	302	205	8.6:1	2 BC

RANCHERO RATING PLATE

The information indicated on the rating plate is the vehicle identification number, the body type, the exterior color, the interior trim, the date manufactured, the D.S.O., the rear axle and the transmission type.

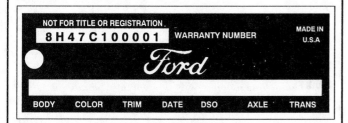

THE VEHICLE IDENTIFICATION NUMBER is a series of letters and numbers on the rating plate. The VIN number identifies the model year, assembly plant, series, engine and production sequence.

FIRST DIGIT: Identifies the model year (1968)

SECOND DIGIT: Identifies the assembly plant

ASSEMBLY PLANT	CODE
Atlanta, GA	A
Oakville, CAN	B
Ontario Truck	C
Dallas, TX	D
Mahwah, NJ	E
Dearborn, MI	F
Chicago, IL	G
Lorain, OH	H
Los Angeles, CA	J
Kansas City, KS	K
Michigan Truck	L
Norfolk, VA	N
Twin Cities, MN	P
San Jose, CA	R
Allen Park, MI (Pilot Plant)	S
Metuchen, NJ	T
Louisville, KY	U
Wayne, MI	W
St. Thomas, CAN	X
Wixom, MI	Y
St. Louis, MO	Z

THIRD AND FOURTH DIGIT: Identify the body serial code

BODY TYPE	CODE
2-Door w/bench	47
2-Door w/bench	48
2-Door w/bucket	48
2-Door w/bucket	49

FIFTH DIGIT: Identifies the engine

ENGINE	CODE
200 cid, 6 cyl.	T
302 cid, 8 cyl.	F
302 cid, 8 cyl.	J
390 cid, 8 cyl.	Y
390 cid, Prem. Fuel	X
390 cid, GT	S

LAST SIX DIGITS: Identify the consecutive unit number

THE VEHICLE DATA appears on the line following the vehicle identification number.

THE BODY CODE indicates the body type.

BODY TYPE	CODE
2-Door	*66A
2-Door (1)	*66B
2-Door (2)	**66B
2-Door (2)	**66D

* Bench seat
** Bucket seat

THE EXTERIOR COLOR CODE indicates the paint color used on the veicle.

COLOR	CODE
Raven Black	A
Royal Maroon	B
Gulf Stream Aqua	F
Lime Gold	I
Wimbledon White	M
Diamond Blue	N
Seafoam Green	O
Brittany Blue	Q
Highland Green	R
Candyapple Red	T
Tahoe Turquoise	U
Meadowlark Yellow	W
Presidential Blue	X
Sunlight Gold	Y
Pebble Beige	6

THE INTERIOR TRIM CODE indicates the trim color and material used on the vehicle.

COLOR	VINYL	CLOTH	LEATHER	CODE
Black	•			2A,5A,8A
Black	•			HA
Dk./Lt. Blue	•			2B,5B,HB
Dk. Blue	•			8B
Pastel Parchment	•			2U,5U,8U
Pastel Parchment	•			BU,EU,HU
Pastel Parchment	•			KU,LU
Dk. Red	•			5D,8D,HD

THE TRANSMISSION CODE indicates the transmission type installed in the vehicle.

TYPE	CODE
3-Speed manual	1
4-Speed manual	5
Automatic (C-4)	W
Automatic (C-6)	U

THE REAR AXLE CODE indicates the ratio of the rear axle installed in the vehicle. A number designates a conventional axle, while a letter designates a locking differential.

RATIO	CODE
2.75:1	1
2.79:1	2
2.83:1	4
3.00:1	5
3.20:1	6
3.25:1	7
3.50:1	8
3.10:1	9
3.00:1	E
3.20:1	F
3.25:1	G
3.50:1	H

THE DATE CODE indicates the date the vehicle was manufactured. A number signifying the date precedes the month code letter. A second year code letter will be used if the model exceeds 12 months.

MONTH	FIRST YEAR	SECOND YEAR
January	A	N
February	B	P
March	C	Q
April	D	R
May	E	S
June	F	T
July	G	U
August	H	V
September	J	W
October	K	X
November	L	Y
December	M	Z

THE D.S.O. CODE: Units built on a Domestic Special Order, Foreign Special Order or other special orders will have the complete order number in this space. Also to appear in this space is the two-digit code number of the District which ordered the unit. If the unit is a regular production unit, only the District code number will appear.

DISTRICT	CODE
Boston	11
New York	13
Newark	15
Philadelphia	16
Washington	17
Atlanta	21
Charlotte	22
Jacksonville	24
Richmond	25
Cincinnati	27
Louisville	28
Cleveland	32
Detroit	33
Indianapolis	34
Lansing	35
Buffalo	37
Pittsburgh	38
Chicago	41
Fargo	42
Milwaukee	43
Twin Cities	44
Davenport	45
Denver	51
Des Moines	52
Kansas City	53
Omaha	54
St. Louis	55
Dallas	61
Houston	62
Memphis	63
New Orleans	64
Oklahoma City	65
Los Angeles	71
San Jose	72
Salt Lake City	73
Seattle	74
Phoenix	75
Ford of Canada	81
Government	83
Home Office Reserve	84
American Red Cross	85
Transportation Services	89
Export	90's

ENGINE SPECIFICATIONS:

ENGINE CODE	NO. CYL.	CID	HORSE-POWER	COMP. RATIO	CARB
T	6	200	120	9.2:1	1 BC
F	8	302	210	9.5:1	2 BC
J	8	302	235	10.5:1	4 BC
Y	8	390	265	9.5:1	2 BC
X	8	390	280	10.5:1	2 BC
S	8	390	320	10.5:1	4 BC
W	8	390	325	10.5:1	4 BC

192

1969 BRONCO

1969 CONTACTOR'S SPECIAL

1969 CAMPER SPECIAL

F-SERIES
RATING PLATE

The information indicated on the rating plate is the vehicle identification number, wheelbase, exterior color, model type, body type, transmission type, rear axle, maximum gross vehicle weight (lbs.), certified net horsepower, r.p.m. and D.S.O. numbers.

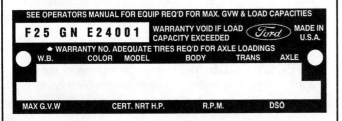

THE VEHICLE IDENTIFICATION NUMBER is a series of letters and numbers on the rating plate. The VIN number identifies the series, engine, assembly plant and the production sequence.

FIRST, SECOND AND THIRD DIGITS: Identify the series

SERIES	CODE
F-100 (4x2)	F10
F-100 (4x4)	F11
F-250 (4x2)	F25
F-250 (4x4)	F26
F-350 (4x2)	F35

FOURTH DIGIT: Identifies the engine

ENGINE	CODE
240 cid, 6 cyl.	A
300 cid, 6 cyl. (LD)	B
390 cid, 8 cyl.	H
360 cid, 8 cyl.	Y

FIFTH DIGIT: Identifies the assembly plant

ASSEMBLY PIANT	CODE
Ontario, CAN	C
Dallas, TX	D
Mawah, NJ	E
Lorain, OH	H
Kansas City, KS	K
Michigan Truck	L
Norfolk, VA	N
Twin Cities, MN	P
San Jose, CA	R
Allen Park, MI (Pilot Plant)	S
Louisville, KY	U

LAST SIX DIGITS: Identify the consecutive unit number

CALENDAR YEAR 1968	NUMBERS
September	D96,000 thru E09,999
October	E10,000 thru E23,999
November	E24,000 thru E37,999
December	E38,000 thru E51,999

CALENDAR YEAR 1969	NUMBERS
January	E52,000 thru E65,999
February	E66,000 thru E79,999
March	E80,000 thru E93,999
April	E94,000 thru F07,999
May	F08,000 thru F21,999
June	F22,000 thru F35,999
July	F36,000 thru F49,999
August	F50,000 thru F63,999

THE VEHICLE DATA appears on the two lines following the vehicle identification number.

THE W.B. (WHEELBASE) CODE indicates the wheelbase in inches.

THE EXTERIOR COLOR CODE indicates the paint color used on the vehicle.

COLOR	CODE
Royal Maroon	B
Pure White	C
Sky View Blue	E
Arcadian Blue	F
Chrome Yellow	G
Cordova	H
Empire	K
Holly Green	L
Wimbledon White	M
Norway Green	N
Boxwood Green Metallic	P
Brittany Blue	Q
Diamond Green	R
Candyapple Red	T
Lunar Green	U
Marlin Blue	X
Reef Aqua	Y
New Lime	2
Twilight Green Metallic	4
Pebble Beige	6
Harbor Blue	7
Prime	9

THE MODEL CODE indicates the model type and gross vehicle weight (lbs.) information.

MODEL	GVW
F-100	5,000
F-101	4,200
F-102*	5,000
F-103	4,500
F-104	4,800
F-110 (4x4)	5,600
F-111 (4x4)	5,000
F-112* (4x4)	5,600
F-113 (4x4)	4,600
F-250	7,500
F-252*	7,500
F-253	6,100
F-254	6,900
F-255*	6,100
F-256*	6,900
F-260 (4x4)	6,800
F-262 (4x4)	7,700
F-263 (4x4)	6,300
F-264* (4x4)	7,700
F-350	8,000
F-351	10,000
F-352*	8,000
F-353	6,600
F-354	8,300
F-355	9,000

* Reference Pennsylvania registration data

THE BODY CODE: The first digit(s) represents the interior trim scheme. The ending two digits are the body type.

THE INTERIOR TRIM CODE indicates the trim color and material used on the vehicle.

COLOR	VINYL	WOVEN PLASTIC	LEATHER	CODE
Lt. Blue	•			B4
Lt. Blue	•	•		2,23,B
Dk. Blue	•			K,K4
Parchment	•			C4
Lt. Parchment	•			L,L4
Lt./Parchment	•	•		33,C
Lt. Parchment w/Black	•	•		3
Black	•			D4,M4,M
Black	•	•		4,43,D
Red	•	•		5,53,E
Dk. Red	•			N

BODY	CODE
Conventional cab	81
Cowl and chassis	84
Cowl and windshield	85

THE TRANSMISSION CODE indicates the transmission type installed on the vehicle.

DESCRIPTION	CODE
4-Speed new process 435	A
3-Speed overdrive T-85	B
3-Speed Ford L.D.	C
3-Speed Warner M.D. T89F	D
3-Speed Warner H.D. T87G	E
4-Speed Syn. Warner T18	F
Automatic	G

THE REAR AXLE CODE:
The first two digits represent the rear axle, the third represents the front axle, if applicable.

RATIO	CODE
4.11 - Ford	05
3.50 - Ford	08
3.70 - Ford	09
3.25 - Ford	17
4.88 - Dana #70	22
4.10 - Dana #60	24
4.56 - Dana #60	25
4.10 - Dana #70	27
4.56 - Dana #70	28
3.73 - Dana #70	36
3.54 - Dana #60	37
3.73 - Dana #60	38
3.25 - Ford	A1*
3.70 - Ford	A2*
4.11 - Ford	A5*
4.10 - Dana #60	B4*
4.56 - Dana #60	B5*
3.50 - Ford	B9*
3.54 - Dana #60	C7*
3.73 - Dana #60	C8*
3.73 - Dana #70	D6*
4.10 - Dana #70	D7*
4.56 - Dana #70	D8*
4.10 - Dana #60-3	E5*
3.54 - Dana #60-3	E9*

* Locking

FRONT AXLE

CAPACITY	CODE
2,500 (limited slip)	J
3,500	K

THE MAX. G.V.W. LBS CODE
indicates the maximum gross vehicle weight in pounds.

THE CERT. NET. H.P. CODE
indicates the certified net horsepower at specified rpm.

THE R.P.M. CODE
indicates the rpm required to develop the certified net horsepower.

THE D.S.O. CODE:
If the vehicle is built on a Direct Special Order, the complete order number will be reflected under the DSO space including the District Code Number.

DISTRICT	CODE
Boston	11
New York	13
Newark	15
Philadelphia	16
Washington	17
Atlanta	21
Charlotte	22
Jacksonville	24
Richmond	25
Cincinnati	27
Louisville	28
Cleveland	32
Detroit	33
Indianapolis	34
Lansing	35
Buffalo	37
Pittsburgh	38
Chicago	41
Milwaukee	43
Twin Cities	44
Davenport	45
Denver	51
Kansas City	53
Omaha	54
St. Louis	55
Dallas	61
Houston	62
Memphis	63
New Orleans	64
Oklahoma City	65
Los Angeles	71
San Jose	72
Salt Lake City	73
Seattle	74
Phoenix	75
Government	83
Home Office Reserve	84
American Red Cross	85
Transportation	89
Export	90's

FORD OF CANADA

	CODE
Central	B1
Eastern	B2
Atlantic	B3
Midwestern	B4
Western	B6
Pacific	B7
Export	I1 thru I7

ENGINE SPECIFICATIONS

ENGINE CODE	NO. CYL.	CID	HORSE-POWER	COMP. RATIO	CARB
A	6	240	150	9.2:1	1 BC
B	6	300	165	8.8:1	1 BC
Y	8	360	215	8.4:1	2 BC
H	8	390	255	8.6:1	2 BC

BRONCO
RATING PLATE

The information indicated on the rating plate is the vehicle identification number, the wheelbase, the exterior color, the model type, the body type, the transmission type, the rear axle, the maximum gross vehicle weight (lbs.), the certified net horsepower, the r.p.m. and the D.S.O. numbers.

THE VEHICLE IDENTIFICATION NUMBER is a series of letters and numbers on the rating plate. The VIN number identifies the series, engine, assembly plant and the production sequence.

FIRST, SECOND AND THIRD DIGITS: Identify the series

SERIES	CODE
U-100 pickup	U14
U-100 wagon	U15

FOURTH DIGIT: Identifies the engine

ENGINE	CODE
170 cid, 6 cyl.	F
302 cid, 8 cyl.	G

FIFTH DIGIT: Identifies the assembly plant

ASSEMBLY PLANT	CODE
Lorain, OH	H
Michigan Truck	L
San Jose, CA	R
Allen Park, MI (Pilot Plant)	S

LAST SIX DIGITS: Identify the consecutive unit number

CALENDAR YEAR - 1968	NUMBERS
August - 1969 Model	D82,000 thru D94,999
September	D96,000 thru E09,999
October	E10,000 thru E23,999
November	E24,000 thru E37,999
December	E38,000 thru E51,999

CALENDAR YEAR - 1969	NUMBERS
January	E52,000 thru E65,999
February	E66,000 thru E79,999
March	E80,000 thru E93,999
April	E94,000 thru F07,999
May	F08,000 thru F21,999
June	F22,000 thru F35,999
July	F36,000 thru F49,999
August	F50,000 thru F63,999

THE VEHICLE DATA appears on the two lines following the vehicle identification number.

THE W.B. (WHEELBASE) CODE indicates the wheelbase in inches.

THE EXTERIOR COLOR CODE indicates the paint color used on the vehicle.

COLOR	CODE
Raven Black	A
Royal Maroon	B
Sky View Blue	E
Chrome Yellow	G
Cordova	H
Empire Green	K
Wimbledon White	M
Norway Green	N
Boxwood Green	P
Candyapple Red	T
Lunar Green	U
Reef Aqua	Y
New Lime	2
Pebble Beige	6
Harbor Blue	7
Prime	9

THE MODEL CODE indicates the model type.

TYPE	CODE
Sports utility	U-140
Heavy duty	U-142
Wagon	U-150
Heavy duty	U-152

THE BODY CODE: The letter and numerals under BODY designate the interior trim and body type (the letter identifies the interior trim scheme and the numerals identify the body or cab type.

THE INTERIOR TRIM CODE indicates the key to the trim color and material used on the vehicle.

COLOR	VINYL	CLOTH	LEATHER	CODE
Lt. Parchment	•			3
Pastel Parchment	•			9,94,9U

BODY	CODE
Open body (roadster)	96
Sports utility	97
Long roof (wagon)	98

THE TRANSMISSION CODE indicates the transmission type installed in the vehicle.

TYPE	CODE
3-Speed Ford L.D.	C

THE REAR AXLE CODE indicates the ratio of the rear axle installed in the vehicle.

RATIO	CODE
4.11	03
4.57	04
4.11	05
4.57	06
3.50	08
3.50	18
4.11	A3
4.11	A5
3.50	B8
3.50	B9

THE MAX. G.V.W. LBS. CODE indicates the maxiumum gross vehicle weight in pounds.

THE CERT. NET. H.P. CODE indicates the certified net horsepower at specified rpm.

THE D.S.O. CODE: If the vehicle is built on Direct Special Order the complete order number will be reflected under the DSO space, including the District Code Number.

DISTRICT	CODE
Boston	11
New York	13
Newark	15
Philadelphia	16
Washington	17
Atlanta	21
Charlotte	22
Jacksonville	24
Richmond	25
Cincinnati	27
Louisville	28
Cleveland	32
Detroit	33
Indianapolis	34
Lansing	35
Buffalo	37
Pittsburgh	38
Chicago	41
Milwaukee	43
Twin Cities	44
Davenport	45
Denver	51
Kansas City	53
Omaha	54
St. Louis	55
Dallas	61
Houston	62
Memphis	63

ENGINE SPECIFICATIONS:

ENGINE CODE	NO. CYL.	CID	HORSE-POWER	COMP. RATIO	CARB
F	6	170	100	8.7:1	1 BC
G	8	302	205	8.6:1	2 BC

RANCHERO
RATING PLATE

The information indicated on the rating plate is the vehicle identification number, the body type, the exterior color, the interior trim, the date manufactured, the D.S.O., the rear axle and the transmission type.

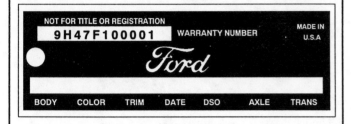

NOT FOR TITLE OR REGISTRATION

9H47F100001   WARRANTY NUMBER

MADE IN U.S.A

Ford

| BODY | COLOR | TRIM | DATE | DSO | AXLE | TRANS |

THE VEHICLE IDENTIFICATION NUMBER is a series of letters and numbers on the rating plate. The VIN number identifies the model year, assembly plant, series, engine and production sequence.

FIRST DIGIT: Identifies the model year (1969)

SECOND DIGIT: Identifies the assembly plant

ASSEMBLY PLANT	CODE
Atlanta, GA	A
Oakville, CAN	B
Ontario Truck	C
Dallas, TX	D
Mahwah, NJ	E
Dearborn, MI	F
Chicago, IL	G
Lorain, OH	H
Los Angeles, CA	J
Kansas City, KS	K
Michigan Truck	L
Norfolk, VA	N
Twin Cities, MN	P
San Jose, CA	R
Allen Park, MI	S
Metuchen, NJ	T
Louisville, KY	U
Wayne, MI	W
St. Thomas, CAN	X
Wixom, MI	Y
St. Louis, MO	Z

THIRD AND FOURTH DIGITS: Identify the body serial code

BODY	CODE
Ranchero	47
Ranchero 500	48
Ranchero GT	49

FIFTH DIGIT: Identifies the engine

ENGINE	CODE
250 cid, 6 cyl.	L
302 cid, 8 cyl.	F
351 cid, 8 cyl.	H
351 cid, 8 cyl.	M
390 cid, 8 cyl.	Y
428 cid, 8 cyl. (CJ)	Q
428 cid, 8 cyl. (CJ) (RAM AIR)	R

LAST SIX DIGITS: Identify the consecutive unit number

THE VEHICLE DATA appears on the line following the vehicle identification number.

THE BODY CODE indicates the body type.

TYPE	CODE
Ranchero	*66A
Ranchero 500	*66B
Ranchero 500 OPT.	**66B
Ranchero GT	*66C
Ranchero GT	**66D

* Bench seat
** Bucket seats

THE EXTERIOR COLOR CODE indicates the paint color used on the vehicle.

COLOR	CODE
Raven Black	A
Royal Maroon	B
Black Jade	C
Aztec Aqua	E
Gulfstream Aqua	F
Lime Gold	I
Wimbledon White	M
Brittany Blue	Q
Champagne Gold	S
Candyapple Red	T
Meadowlark Yellow	W
Presidential Blue	X
Indian Fire	Y
New Lime	2
Dresden Blue	8

THE INTERIOR TRIM CODE indicates the trim color and material used on the vehicle.

COLOR	CODE
Black vinyl	2A,6A,8A
Black vinyl	HA,QA,VA
Black vinyl	WA
Black cloth/vinyl	5A,EA
Lt. Blue vinyl	5B,HB
Dk./Lt. Blue vinyl	2B,6B,8B
Lt. Nugget Gold vinyl	2Y
Dk. Red vinyl	5D,6D,8D
Dk. Red vinyl	HD
Lt. Nugget Gold vinyl	6Y,8Y,HY
White cloth/vinyl	EW
White vinyl	QW,VW,WW

THE DATE CODE indicates the date the vehicle was manufactured. The number indicates the day of the month and the letter indicates the month. A second year code letter will be used if the model exceeds 12 months.

MONTH	FIRST YEAR	SECOND YEAR
January	A	N
February	B	P
March	C	Q
April	D	R
May	E	S
June	F	T
July	G	U
August	H	V
September	J	W
October	K	X
November	L	Y
December	M	Z

THE REAR AXLE CODE indicates the ratio of the rear axle installed in the vehicle.

RATIO	CONVENTIONAL	LIMITED-SLIP
3.00:1	6	O
3.25:1	9	R
3.50:1	A	S
3.91:1	—	V

THE TRANSMISSION CODE indicates the transmission type installed in the vehicle.

TYPE	CODE
3-Speed manual	1
4-Speed manual - wide ratio	5
4-Speed manual - close ratio	6
Automatic (C4)	W
Automatic (C6)	U
Automatic (MX)	Y
Automatic (FMX)	X
FMX-L	T
Automatic (C6 Special)	Z

THE D.S.O. CODE: Units built on a Domestic Special Order, Foreign Special Order, or other special orders will have the complete order number in this space. Also to appear in this space is the two-digit code number of the District which ordered the unit. If the unit is a regular production unit, only the District code number will appear.

DISTRICT	CODE
Boston	11
New York	13
Newark	15
Philadelphia	16
Washington	17
Atlanta	21
Charlotte	22
Jacksonville	24
Richmond	25
Louisville	28
Cleveland	32
Detroit	33
Lansing	35
Buffalo	37
Pittsburgh	38
Chicago	41
Milwaukee	43
Twin Cities	44
Indianapolis	46
Cincinnati	47
Denver	51
Kansas City	53
Omaha	54
St. Louis	55
Davenport	56
Dallas	61
Houston	62
Memphis	63
New Orleans	64
Oklahoma City	65
Los Angeles	71
San Jose	72
Salt Lake City	73
Seattle	74
Phoenix	75
Government	83
Home Office Reserve	84
American Red Cross	85
Transportation Services	89
Export	90's

ENGINE SPECIFICATIONS:

ENGINE CODE	NO. CYL.	CID	HORSE-POWER	COMP. RATIO	CARB
L	6	250	155	9.0:1	1 BC
F	8	302	210	9.5:1	2 BC
H	8	351	250	9.5:1	2 BC
M	8	351	290	10.7:1	4 BC
Y	8	390	320	10.5:1	4 BC
Q	8	428	335	10.6:1	4 BC
R	8	428	RAM AIR	10.6:1	4 BC

1970 F-100 PICKUP

1970 RANCHERO

1970 RANGER XLT

1970 RANCHERO

F-SERIES RATING PLATE

The information indicated on the rating plate is the vehicle identification number, the wheelbase, exterior color, model type, body type, transmission type, rear axle, maximum gross vehicle weight in lbs., certified net horsepower, r.p.m. and D.S.O. numbers.

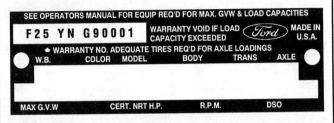

THE VEHICLE IDENTIFICATION NUMBER is a series of letters and numbers on the rating plate. The VIN number identifies the series, engine, assembly plant and the production sequence.

FIRST, SECOND AND THIRD DIGITS: Identify the series

SERIES	CODE
F-100 (4x2)	F10
F-100 (4x4)	F11
F-250 (4x2)	F25
F-250 (4x4)	F26
F-350 (4x2)	F35

FOURTH DIGIT: Identifies the engine

ENGINE	CODE
240 cid, 6 cyl.	A
300 cid, 6 cyl.	B
302 cid, 8 cyl.	G
390 cid, 8 cyl.	H
360 cid, 8 cyl.	Y

FIFTH DIGIT: Identifies the assembly plant

ASSEMBLY PLANT	CODE
Ontario, CAN	C
Dallas, TX	D
Mawah, NJ	E
Lorain, OH	H
Kansas City, KS	K
Michigan Truck	L
Norfolk, VA	N
Twin Cities, MN	P
San Jose, CA	R
Allen Park, MI (Pilot Plant)	S
Louisville, KY	U
Kentucky Truck	V

LAST SIX DIGITS: Identify the consecutive unit number

CALENDAR YEAR - 1969	NUMBERS
August - 1970 model	G30,000 thru G49,999
September	G50,000 thru G69,999
October	G70,000 thru G89,999
November	G90,000 thru H09,999
December	H10,000 thru H29,999

CALENDAR YEAR - 1970	NUMBERS
January	H30,000 thru H49,999
February	H50,000 thru H69,999
March	H70,000 thru H89,999
April	H90,000 thru J09,999
May	J10,000 thru J29,999
June	J30,000 thru J49,999
July	J50,000 thru J69,999
August	J70,000 thru J89,999

THE VEHICLE DATA appears on the two lines following the vehicle identification number.

THE W.B. (WHEELBASE) CODE indicates the wheelbase in inches.

THE EXTERIOR COLOR CODE indicates the paint colors used on the vehicle.

COLOR	CODE
Raven Black	A
Royal Maroon	B
Pure White	C
Pinto Yellow	D
Sky View Blue	E
Mohave Tan	F
Chrome Yellow	G
Cactus Green	H
Lime Metallic	I
Red	J
Wimbledon White	M
Norway Green	N
Boxwood Green	P
Brittany Blue Metallic	Q
Yucatan Gold	R
Champagne Gold Metallic	S
Candyapple Red	T
Tampico Yellow	W
Reef Aqua	Y
Baja Beige	Z
New Lime	2
Crystal Green	4
Diamond Blue	5
Acapulco Blue Metallic	6
Harbor Blue	7
Morning Gold	8

THE MODEL CODE indicates the model type and gross vehicle weight (lbs.) information.

F-100 SERIES

MODEL	GVW
F-100	5,000
F-101	4,200
F-102*	5,000
F-103	4,500
F-104	4,800

F-100 SERIES (4x4)

MODEL	GVW
F-110	5,600
F-111	5,000
F-112*	5,600
F-113	4,600

F-250 SERIES

MODEL	GVW
F-250	7,500
F-252*	7,500
F-253	6,100
F-254	6,900
F-255*	6,100
F-256*	6,900

F-250 SERIES (4x4)

MODEL	GVW
F-260	6,800
F-262	7,700
F-263	6,300
F-264*	7,700

F-350 SERIES (4x4)

MODEL	GVW
F-350	8,000
F-351	10,000
F-352*	8,000
F-353	6,600
F-354	8,300
F-355	9,000

* Reference Pennsylvania registration data

THE BODY CODE: The first digit represents the interior trim scheme, second is seat type and third is body code.

THE INTERIOR TRIM CODE indicates the key to the trim color and material used on the vehicle.

COLOR	VINYL	CLOTH	LEATHER	CODE
Lt./Med. Blue	•			2,1B,23
Lt./Med. Blue	•			B,K,GB,BB
Lt./Med. Blue	•			7B,AB
Med. Blue	•	•		S
Parchment	•			3,1U,AA
Parchment	•			C,L,GU
Parchment	•			BU,7U,AU
Parchment	•	•		T
Parchment	•			C4,L4,T4
Parchment	•			SU,EU,9U
Black	•			4,1A,4B
Black	•			4C,DB,4A
Black	•			43,D,M
Black	•			GA,BA,7A
Black	•	•		U
Black	•			FA

COLOR	VINYL	CLOTH	LEATHER	CODE
Red/Dk. Red	•			5,1D,53
Red/Dk. Red	•			E,N,GD
Red/Dk. Red	•			BD,7D,AD
Dk. Red	•	•		V
Lt./Med. Ivy Green		•		6,1G,63
Lt./Med. Ivy Green		•		F,O,GG
Lt./Med. Ivy Green		•		BG,7G,AG
Lt./Med. Ivy Green		•		AU
Med. Ivy Green	•	•		W
Black/Blue	•			2B,2C,BB
Black/Blue	•			4A
Black/Parchment	•			3B,3C,CB
Black/Parchment	•			4A
Black/Red	•			5B,5C,EB
Black/Red	•			4A
Black/Ivy	•			6B,6C,FB
Black/Ivy	•			4A

THE TRANSMISSION CODE indicates the transmission type installed in the vehicle.

TYPE	CODE
4-Speed new process 435	A
3-Speed overdrive T-85	B
3-Speed Ford L.D.	C
3-Speed Warner M.D. T89F	D
3-Speed Warner H.D. T87G	E
4-Speed Syn. Warner T18	F
Automatic	G
4-Speed Warner T19	P

THE REAR AXLE CODE: The first two digits represent the rear axle and the third represents the front axle, if applicable.

REAR AXLE CODES

DESCRIPTION	RATIO	LBS.	CODE
Ford	4.11	3,300	05
Ford	4.57	3,300	06
Ford	3.50	3,300	08
Ford	3.70	3,300	09
Ford	3.25	3,050	10

	RATIO	LBS.	CODE
Dana #70	4.88	7,400	22
Dana #60	4.10	5,200	24
Dana #60	4.56	5,200	25
Dana #70	4.10	7,400	27
Dana #70	4.56	7,400	28
Ford	3.25	3,300	A1*
Ford	3.70	3,300	A2*
Ford	4.11	2,780	A3*
Ford	4.11	3,300	A5*
Dana #60	4.10	5,200	B4*
Dana #60	4.56	5,200	B5*
Ford	3.50	3,300	B9*
Dana #60	3.54	5,200	C7*
Dana #60	3.73	5,200	C8*
Dana #70	4.88	7,400	D2*
Dana #70	3.73	7,400	D6*
Dana #70	4.10	7,400	D7*
Dana #70	4.56	7,400	D8*
Ford	3.25	—	H1*
Ford	3.50	—	H2*
Ford	4.09	—	H3*

* Locking

FRONT AXLE CODES

LBS.	CODE
2,500	J
3,500	K

THE MAX. G.V.W. LBS. CODE indicates the maximum gross vehicle weight in pounds.

THE CERT. NET H.P. CODE indicates the certified net horsepower at specified rpm.

THE R.P.M. CODE indicates the specified rpm required to develop the certified net horsepower.

THE D.S.O. CODE: If the vehicle is built on a Direct Special order the complete order number will be reflected under the D.S.O. space including the District Code Number.

DISTRICT	CODE
Boston	11
New York	13
Newark	15
Philadelphia	16
Washington	17
Atlanta	21
Charlotte	22
Jacksonville	24
Richmond	25
Louisville	28
Cleveland	32
Detroit	33
Lansing	35
Buffalo	37
Pittsburgh	38
Chicago	41
Milwaukee	43
Twin Cities	44
Indianapolis	46
Cincinnati	47
Denver	51
Kansas City	53
Omaha	54
St. Louis	55
Davenport	56
Dallas	61
Houston	62
Memphis	63
New Orleans	64
Oklahoma City	65
Los Angeles	71
San Jose	72
Salt Lake City	73
Seattle	74
Phoenix	75
Government	83
Home Office Reserve	84
American Red Cross	85
Body Company	87
Transportation Services	89
Export	90's

FORD OF CANADA

Central	B1
Eastern	B2
Atlantic	B3
Midwestern	B4
Western	B6
Pacific	B7
Export	I2

ENGINE SPECIFICATIONS:

ENGINE CODE	NO. CYL.	CID	HORSE-POWER	COMP. RATIO	CARB
A	6	240	150	9.2:1	1 BC
B	6	300	165	8.8:1	1 BC
G	8	302	205	8.6:1	2 BC
H	8	360	215	8.4:1	2 BC
Y	8	390	255	8.6:1	2 BC

BRONCO RATING PLATE

The information indicated on the rating plate is the vehicle identification number, the wheelbase, the exterior color, the model type, the body type, the transmission type, the rear axle, the maximum gross vehicle weight (lbs.), the certified net horsepower, the r.p.m. and the D.S.O. numbers.

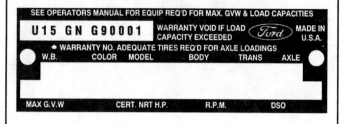

THE VEHICLE IDENTIFICATION NUMBER is a series of letters and numbers on the rating plate. The VIN number identifies the series, engine, assembly plant and the production sequence.

FIRST, SECOND AND THIRD DIGITS: Identify the series

SERIES	CODE
U-100 Pickup	U14
U-100 Wagon	U15

FOURTH DIGIT: Identifies the engine

ENGINE	CODE
170 cid, 6 cyl.	F
302 cid, 8 cyl.	G

FIFTH DIGIT: Identifies the assembly plant

ASSEMBLY PLANT	CODE
Lorain, OH	H
Michigan Truck	L
San Jose, CA	R
Allen Park, MI (Pilot Plant)	S

LAST SIX DIGITS: Identify the consecutive unit number

CALENDAR YEAR - 1969	NUMBERS
August - 1970 model	G30,000 thru G49,999
September	G50,000 thru G69,999
October	G70,000 thru G89,999
November	G90,000 thru H09,999
December	H10,000 thru H29,999

CALENDAR YEAR - 1970	NUMBERS
January	H30,000 thru H49,999
February	H50,000 thru H69,999
March	H70,000 thru H89,999
April	H90,000 thru J09,999
May	J10,000 thru J29,999
June	J30,000 thru J49,999
July	J50,000 thru J69,999
August	J70,000 thru J89,999

THE VEHICLE DATA appears on the two lines following the vehicle identification number.

THE W.B. (WHEELBASE) CODE indicates the wheelbase in inches.

THE EXTERIOR COLOR CODE indicates the paint colors used on the vehicle.

COLOR	CODE
Raven Black	A
Royal Maroon	B
Pinto Yellow	D
Sky View Blue	E
Mojave Tan	F
Chrome Yellow	G
Carmel Bronze Metallic	K
Wimbledon White	M
Norway Green	N
Boxwood Green	P
Yucatan Gold	R
Candyapple Red	T
Reef Aqua	Y
New Lime	2
Diamond Blue	5
Acapulco Blue	6
Harbor Blue	7
Prime	9

THE MODEL CODE indicates the model type and the gross vehicle weight (lbs.) information.

MODEL	GVW	TYPE
U-140	3,900	Pickup
U-142	4,700	Heavy Duty
U-150	3,900	Wagon
U-152	4,700	Heavy Duty

THE BODY CODE: The letter and numerals under body designate the interior trim and body type (the letter identifies the interior trim scheme and the numerals identify the body or cab type).

CAB TRIM

TYPE	COLOR	CODE
Vinyl standard bench	Parchment	3
Vinyl optional bench	Parchment	9
Vinyl bucket option	Parchment	94

BODY	CODE
Open body (roadster)	96
Sports utility	97
Long roof (wagon)	98

THE TRANSMISSION CODE indicates the transmission type used in the vehicle.

TYPE	CODE
3-Speed Ford L.D.	C

THE REAR AXLE CODE indicates the ratio of the rear axle installed in the vehicle.

AXLE	CODE
4.11	03
4.57	04
4.11	05
4.57	06
3.50	08
3.50	18
4.11	A3
4.11	A5
3.50	B8
3.50	B9

THE MAX. G.V.W. LBS. CODE indicates the maximum gross vehicle weight in pounds.

THE CERT. NET H.P. CODE indicates the certified net horsepower at specified rpm.

THE R.P.M. CODE indicates the specified rpm required to develop the certified net horsepower.

THE D.S.O. CODE: If the vehicle is built on a Direct Special Order the complete order number will be reflected under the D.S.O. space including the District Code Number.

DISTRICT	CODE
Boston	11
New York	13
Newark	15
Philadelphia	16
Washington	17
Atlanta	21
Charlotte	22
Jacksonville	24
Richmond	25
Louisville	28
Cleveland	32
Detroit	33
Lansing	35
Buffalo	37
Pittsburgh	38
Chicago	41
Milwaukee	43
Twin Cities	44
Indianapolis	46
Cincinnati	47
Denver	51
Kansas City	53
Omaha	54
St. Louis	55
Davenport	56
Dallas	61
Houston	62
Memphis	63

New Orleans	64
Oklahoma City	65
Los Angeles	71
San Jose	72
Salt Lake City	73
Seattle	74
Phoenix	75
Government	83
Home Office Reserve	84
American Red Cross	85
Body Company	87
Transportation Services	89
Export	90's

FORD OF CANADA

Central	B1
Eastern	B2
Atlantic	B3
Midwestern	B4
Western	B6
Pacific	B7
Export	I2

ENGINE SPECIFICATIONS:

ENGINE CODE	NO. CYL.	CID	HORSE-POWER	COMP. RATIO	CARB
F	6	170	100	8.7:1	1 BC
G	8	302	205	8.6:1	2 BC

RANCHERO
RATING PLATE

The information indicated on the rating plate is the vehicle identification number, the body type, the exterior color, the interior trim, the rear axle, the transmission type and the D.S.O. numbers.

VEH. IDENT NO.	BODY	COL.
OH47F100001		

TRIM	AXLE	TRANS	DSO

NOT FOR TITLE OR REGISTRATION

MADE IN U.S.A

THE VEHICLE IDENTIFICATION NUMBER is a series of letters and numbers on the rating plate. The VIN number identifies the model year, the assembly plant, the series, the engine and the production sequence.

FIRST DIGIT: Identifies the model year 1970

SECOND DIGIT: Identifies the assembly plant

ASSEMBLY PLANT	CODE
Atlanta, GA	A
Oakville, CAN	B
Dallas, TX	D
Mahwah, NJ	E
Dearborn, MI	F
Chicago, IL	G
Lorain, OH	H
Los Angeles, CA	I
Kansas City, KS	K
Norfolk, VA	N
Twin Cities, MN	P
San Jose, CA	R
Allen Park, MI	S
Metuchen, NJ	T
Louisville, KY	U
Wayne, MI	W
St. Thomas, CAN	X
Wixom, MI	Y

THIRD AND FOURTH DIGITS: Identify the body serial code

The two-digit numeral which follows the assembly plant code identifies the body series. This two-digit number is used in conjunction with the body style code, in the vehicle data, which consists of a two-digit number with a letter suffix. The following chart lists the body serial codes, body style codes and the model.

SERIES	CODE
Ranchero	47
Ranchero GT	48
Ranchero Squire	49

FIFTH DIGIT: Identifies the engine

ENGINE	CODE
250 cid, 6 cyl.	L
250 cid, 6 cyl.	*3
302 cid, 8 cyl.	F
302 cid, 8 cyl.	*6
351 cid, 8 cyl.	H
351 cid, 8 cyl.	M
429 cid, 8 cyl.	N
429 cid, 8 cyl. (CJ)	C
429 cid, 8 cyl. (CJ Ram Air)	J

* Low compression
** Premium fuel
+ Ram air induction

SIXTH DIGIT: Identifies the consecutive unit number

THE VEHICLE DATA appears on the two lines following the vehicle identification number.

THE BODY CODE indicates the body type.

BODY	CODE
Ranchero	66A
Ranchero 500	66B
Ranchero GT	66C
Ranchero Squire	66E

THE EXTERIOR COLOR CODE indicates the paint colors used on the vehicle.

COLOR	CODE
Black	A
Dk. Maroon	B
Dk. Ivy Green Metallic	C
Competition Yellow	D
Bright Gold Metallic	K
White	M
Pastel Blue	N
Med. Ivy Green Metallic	P
Med. Blue Metallic	Q
Med. Gold Metallic	S
Red	T
Dk. Blue	X
Grabber Green	Z
Vermilion	1
Lt. Ivy Yellow	2
Lt. Gold	8

GLAMOUR PAINTS - OPTIONAL

COLOR	CODE
Ivy Bronze Metallic - Green Fire - Green Stardust	19
Olive Bronze Metallic - Olive Fire - Olive Stardust	09
Fall Bronze Metallic - Bronze Fire - Bronze Stardust	89
Med. Red Metallic - Burgundy Fire - Red Stardust	59

THE INTERIOR TRIM CODE indicates the key to the trim color and material used on the vehicle.

COLOR	CODE
Black vinyl	GA,UA
Black knit/vinyl	TA
Med. Blue vinyl	GB,UB
Med. Blue knit/vinyl	QB,TB
Dk. Red vinyl	GD
Dk. Red knit/vinyl	QD,TD
Med. Ginger vinyl	GF
White vinyl w/Black	GW
White knit/vinyl w/Black	QW
Black knit/vinyl	QA
Med./Lt. Ivy Green knit/vinyl	TG
Lt. Gold vinyl	UY
Lt. Nugget Gold knit/vinyl	TY

THE TRANSMISSION CODE indicates the transmission type installed in the vehicle.

TYPE	CODE
3-Speed manual	1
4-Speed manual - wide ratio	5
4-Speed manual - close ratio	6
Semi-automatic stick shift	V
Automatic (C4)	W
Automatic (C6)	U
Automatic (FMX)	X
Automatic (C6 Special)	Z

THE REAR AXLE CODE indicates the ratio of the rear axle installed in the vehicle.

RATIO	CONVENTIONAL	LIMITED-SLIP
3.00:1	6	O
3.25:1	9	K
3.91:1	—	V

THE D.S.O. CODE: Units built on a Domestic Special Order, Foreign Special Order, or other special orders will have the complete order number in this space. Also to appear in this space is the two-digit code number of the District which ordered the unit. If the unit is a regular production unit, only the District code number will appear.

DISTRICT	CODE
Boston	11
New York	13
Newark	15
Philadelphia	16
Washington	17
Atlanta	21
Charlotte	22
Jacksonville	24
Richmond	25
Louisville	28
Cleveland	32
Detroit	33
Lansing	35
Buffalo	37
Pittsburgh	38
Chicago	41
Milwaukee	43
Twin Cities	44
Indianapolis	46
Cincinnati	47
Denver	51
Kansas City	53
Omaha	54
St. Louis	55
Davenport	56
Dallas	61
Houston	62
Memphis	63
New Orleans	64
Oklahoma City	65
Los Angeles	71
San Jose	72
Salt Lake City	73
Seattle	74
Phoenix	75
Government	83
Home Office Reserve	84
American Red Cross	85
Transportation Services	89
Export	90's

ENGINE SPECIFICATIONS:

ENGINE CODE	NO. CYL.	CID	HORSE-POWER	COMP. RATIO	CARB
L	6	250	155	9.1:1	1 BC
F	8	302	210	9.5:1	2 BC
H	8	351	250	9.5:1	2 BC
M	8	351	300	11.0:1	4 BC
C	8	429	345	11.5:1	4 BC
N	8	429	375	10.5:1	4 BC

1971 RANCHERO GT

1971 BRONCO

1971 F-250 PICKUP

F-SERIES
RATING PLATE

The information indicated on the rating plate is the vehicle identification number, the wheelbase, exterior color, model type, body type, transmission type, rear axle, maximum gross vehicle weight (lbs.), certified net horsepower, r.p.m. and D.S.O. numbers.

VEHICLE IDENTIFICATION NUMBER

The first series of letters and numbers on the rating plate, the VIN number identifies the series, engine, assembly plant and the production sequence.

FIRST, SECOND AND THIRD DIGITS: Identify the series

SERIES	CODE
F-100 (4x2)	F10
F-100 (4x4)	F11
F-250 (4x2)	F25
F-250 (4x4)	F26
F-350 (4x2)	F35

FOURTH DIGIT: Identifies the engine

ENGINE	CODE
302 cid, 8 cyl.	G
240 cid, 6 cyl.	A
360 cid, 6 cyl.	B
360 cid, 8 cyl.	Y
390 cid, 8 cyl.	H
300 cid (600 series), 6 cyl.	B
330 cid, 8 cyl.	C
330 cid, 8 cyl.	D
361 cid, 8 cyl.	E
240 cid, 6 cyl.	1
300 cid, 6 cyl.	2
360 cid, 8 cyl.	8

FIFTH DIGIT: Identifies the assembly plant

ASSEMBLY PLANT	CODE
Ontario Truck	C
Mahwah, NJ	E
Lorain, OH	H
Kansas City, KS	K
Michigan Truck	L
Norfolk, VA	N
Twin Cities, MN	P
San Jose, CA	R
Allen Park, MI	S
Kentucky Truck	V

LAST SIX DIGITS: Identify the consecutive unit number

CALENDAR YEAR - 1970	NUMBERS
July 1970 Model	J50,000 thru J69,999
July 1971 Model	J70,000 thru J79,999
August	J80,000 thru J99,999
September	K00,000 thru K19,999
October	K20,000 thru K39,999
November	K40,000 thru K59,999
December	K60,000 thru K79,999

CALENDAR YEAR - 1971	NUMBERS
January	K80,000 thru K99,999
February	L00,000 thru L19,999
March	L20,000 thru L39,999
April	L40,000 thru L75,999
May	L60,000 thru L79,999
June	L80,000 thru L99,999
July	M00,000 thru M19,999
August	M20,000 thru M29,999

THE VEHICLE DATA appears on the two lines following the vehicle identification number.

THE W.B. (WHEELBASE) CODE indicates the wheelbase in inches.

THE EXTERIOR COLOR CODE indicates the paint color of the vehicle.

COLOR	CODE
Raven Black	A
Pure White	C
Sky View Blue	E
Mohave Tan	F
Chrome Yellow	G
Boxwood Green	K
Wimbledon White	M
Diamond Blue	N
Seafoam Green	O
Candyapple Red	T
Mallard Green	V
Calypso Coral	1
Prairie Yellow	2
Swiss Aqua	3
Regis Red	4
Bahama Blue	6
Prime	9

THE MODEL CODE indicates the model type and the gross vehicle weight (lbs.) information.

F-100 SERIES

PICKUPS	GVW
F-100	5,000
F-101	4,200
F-102*	5,000
F-103	4,500
F-104	4,800

F-100 (4x4) SERIES

F-110	5,600
F-111	5,000
F-112*	5,600
F-113	4,600

F-250 SERIES

F-250	7,500
F-252*	7,500
F-253	6,100
F-254	6,900
F-255*	6,100
F-256*	6,900

F-250 (4x4) SERIES

F-260	6,800
F-262	7,700
F-263	6,300
F-264*	7,700

F-350 SERIES

F-350	8,000
F-351	10,000
F-352*	8,000
F-353	6,600
F-354	8,300
F-355	9,000

* Reference Pennsylvania registration data

THE BODY CODE: The first digit is the cab trim, the second is the seat type and the third is the body code.

CUSTOM	RANGER	RANGER XLT	STD.	CAB/TRIM
A	—	—	1	Explorer Blue
B	K	S	2	Blue
C	L	T	3	Parchment
D	M	U	4	Black
E	N	V	5	Red
F	0	W	6	Green
G	—	—	7	Explorer Green
H	—	—	8	Explorer Ginger
—	—	—	9	Parchment

SEAT CODES

HD BLACK VINYL	KNITTED VINYL	DRIVER W/COMP.	SINGLE DRIVER	FULL WIDTH	DESCRIPTION
A	J	—	—	1	Full width
B	K	—	—	2	Full width-custom
—	—	C	3	—	L-S unison
—	—	D	4	—	L-S #675 Bostrom
—	—	E	5	—	Westcoaster
—	—	F	6	—	Bostrom T-Bar National
—	—	G*	7	—	Cush-N-Aire
—	—	H	8	—	Bostrom Level Air

* w/Lear Siegler companion seat

BODY TYPE	CODE
Conventional cab	81
Cowl and chassis	84
Cowl and windshield	85

THE TRANSMISSION CODE indicates the type of transmission installed in the vehicle.

TYPE	CODE
New process 435 4-speed	A
Warner T-85 overdrive 3-speed	B
Ford - manual - 3-speed	C
Warner T-89C 3-speed	D
Warner T-87G 3-speed	E
Warner T-184 4-speed	F
C-6 automatic	G
Warner T-19 4-speed	P

THE REAR AXLE CODE: The first two digits represent the rear axle and the third represents the front axle, if applicable.

REAR AXLE CODES

DANA

5050 NO.	CODE
3.54	71
3.73	72
4.10	73
4.56	74

LIMITED SLIP

DANA 5050 NO.	CODE
3.54	G1
3.73	G2
4.10	G3

3300 NO. FORD	CODE
3.25	17
3.50	08
3.70	09
4.11	05
3.00	02

3300 NO. FORD LIMITED SLIP	CODE
3.25	A1
3.50	B9
4.11	A5
3.70	A2

3600 NO. FORD LIMITED SLIP	CODE
3.25	H1
3.50	H2
4.09	H3

5200 NO. DANA 60	CODE
4.10	24
4.56	25
3.54	37
3.73	38

5200 NO. DANA 60 LIMITED SLIP	CODE
4.10	B4
4.56	B5
3.54	C7
3.73	C8

FRONT AXLE CODES

DESCRIPTION	CODE
2,500# Dana - 30 - locking	J
3,500# Dana - 44F	K

THE MAX. G.V.W. LBS. CODE indicates the maximum gross vehicle weight in pounds.

THE CERT. NET H.P. indicates the certified net horsepower at specified rpm.

THE R.P.M. indicates the specified r.p.m. required to develop the certified net horsepower.

THE D.S.O. CODE: If the vehicle is built on a direct special order the complete order number will be reflected under the D.S.O. space including the District Code Number.

DISTRICT CODES

DISTRICT	CODE
Boston	11
New York	13
Newark	15
Philadelphia	16
Washington	17
Atlanta	21
Charlotte	22
Jacksonville	24
Richmond	25
Louisville	28
Cleveland	32
Detroit	33
Lansing	35
Buffalo	37
Pittsburgh	38
Chicago	41
Milwaukee	43
Twi Cities	44
Indianapolis	46
Cincinnati	47
Denver	51
Kansas City	53
Omaha	54
St. Louis	55
Davenport	56
Dallas	61
Houston	62
Memphis	63
New Orleans	64
Oklahoma City	65
Los Angeles	71
San Jose	72
Salt Lake City	73
Seattle	74
Phoenix	75
Government	83
Home Office Reserve	84
American Red Cross	85
Transportation Services	89
Body Company	87
Export	90's

FORD OF CANADA

Central	B1
Eastern	B2
Atlantic	B3
Midwestern	B4
Western	B6
Pacific	B7
Export	12

ENGINE SPECIFICATIONS:

ENGINE CODE	NO. CYL.	CID	HORSE-POWER	COMP. RATIO	CARB
A	6	240	150	9.2:1	1 BC
B	6	300	165	8.8:1	1 BC
G	8	302	205	8.6:1	2 BC
Y	8	360	215	8.4:1	2 BC
H	8	390	255	8.6:1	2 BC

BRONCO
RATING PLATE

The information indicated on the rating plate is the vehicle identification number, the wheelbase, the exterior color, the model type, the body type, the transmission type, the rear axle, the maximum gross vehicle weight (lbs.), the certified net horsepower and the D.S.O. numbers.

VEHICLE IDENTIFICATION NUMBER

The first series of letters and numbers on the rating plate, the VIN number identifies the series, engine, assembly plant and the production sequence.

FIRST, SECOND AND THIRD DIGITS: Identify the series

SERIES	CODE
Sports utility (U-140)	U14
Sports utility (U-142/HD)	U14
Wagon (U-150)	U15
Wagon (U-152/HD)	U15

FOURTH DIGIT: Identifies the engine

ENGINE	CODE
170 cid, 6 cyl.	F
302 cid, 8 cyl.	G

FIFTH DIGIT: Identifies the assembly plant

ASSEMBLY PLANT	CODE
Lorain, OH	H
Michigan Truck	L
San Jose, CA	R
Allen Park, MI	S

LAST SIX DIGITS: Identify the consecutive unit number

CALENDAR YEAR - 1970	NUMBERS
July - 1970 model	J50,000 thru J69,999
July - 1971 model	J70,000 thru J79,999
August	J80,000 thru J99,999
September	K00,000 thru K19,999
October	K20,000 thru K39,999
November	K40,000 thru K59,999
December	K60,000 thru K79,999

CALENDAR YEAR - 1971	NUMBERS
January	K80,000 thru K99,999
February	L00,000 thru L19,999
March	L20,000 thru L39,999
April	L40,000 thru L59,999
May	L60,000 thru L79,999
June	L80,000 thru L99,999
July	M00,000 thru M19,999
August	M20,000 thru M29,999

THE VEHICLE DATA appears on the two lines following the vehicle identification number.

THE W.B. (WHEELBASE) CODE indicates the wheelbase in inches.

THE EXTERIOR COLOR CODE indicates the paint color on this vehicle.

COLOR	CODE
Raven Black	A
Sky View Blue	E
Mohave Tan	F
Chrome Yellow	G
Boxwood Green	K
Wimbledon White	M
Diamond Blue	N
Seafoam Green	O
Astra Blue Metallic	R
Candyapple Red	T
Mallard Green	V
Grabber Green Metallic	Z
Calypso Coral	1
Prairie Yellow	2
Swiss Aqua	3
Regis Red	4
Bahama Blue	6
Prime	9

THE MODEL CODE indicates the model type.

MODEL	SERIES	TYPE
U-140	U-100	Sports Utility
U-142	HD package	—
U-150	U-100	Wagon
U-152	HD package	—

THE BODY CODE: The letter and numerals under body designate the interior trim and body type (the letter identifies the interior trim scheme and the numerals identify the body or cab type).

BODY TYPE	CODE
Open body (roadster)	96
Sports utility	97
Long roof (wagon)	98

COLOR

CAB/TRIM	CODE
Parchment	9

SEAT CODES

H.D. VINYL	BRONCO REAR SEAT	STD.	SEATS
B	—	2	Bench seat
C	—	3	Foam Cushion
—	U	4	Bucket Seats
			Driver & Passenger

THE TRANSMISSION CODE indicates the type of transmission installed in the vehicle.

TYPE	CODE
3-Speed manual	C

THE REAR AXLE CODE indicates the ratio of the rear axle installed in the vehicle.

AXLE - FORD 2780 No.	CODE
4.11	03
4.11 (lock)	A3
4.57	04
3.50	18
3.50 (lock)	B8

AXLE - FORD 3300 No.	CODE
4.11	05
4.11 (lock)	A5
3.50	08
3.50 (lock)	B9

THE MAX. G.V.W. LBS. CODE indicates the maximum gross vehicle weight in pounds.

THE CERT. NET H.P. CODE indicates the certified net horsepower at specified r.p.m.

THE R.P.M. CODE indicates the specified r.p.m. required to develop the certified net horsepower.

THE D.S.O. CODE: If the vehicle is built on a Direct Special Order the complete order number will be reflected under the D.S.O. space, including the District Code Number.

DISTRICT	CODE
Boston	11
New York	13
Newark	15
Philadelphia	16
Washington	17
Atlanta	21
Charlotte	22
Jacksonville	24
Richmond	25
Louisville	28
Cleveland	32
Detroit	33
Lansing	35
Buffalo	37
Pittsburgh	38
Chicago	41
Milwaukee	43
Twin Cities	44
Indianapolis	46
Cincinnati	47
Denver	51
Kansas City	53
Omaha	54
St. Louis	55
Davenport	56
Dallas	61

Houston	62
Memphis	63
New Orleans	64
Oklahoma City	65
Los Angeles	71
San Jose	72
Salt Lake City	73
Seattle	74
Phoenix	75
Government	83
Home Office Reserve	84
American Red Cross	85
Transportation Services	89
Body Company	87
Export	90's

FORD OF CANADA

Central	B1
Eastern	B2
Atlantic	B3
Midwestern	B4
Western	B6
Pacific	B7
Export	12

ENGINE SPECIFICATIONS:

ENGINE CODE	NO. CYL.	CID	HORSE-POWER	COMP. RATIO	CARB
F	6	170	100	8.7:1	1 BC
G	8	302	205	8.6:1	2 BC

RANCHERO
RATING PLATE

The information indicated on the rating plate is the vehicle identification number, the body type, the exterior color, the interior trim, the rear axle, the transmission type and the D.S.O. numbers.

```
MANUFACTURED BY                        100001
FORD MOTOR COMPANY
   08/70 THIS VEHICLE CONFORMS
   TO ALL APPLICABLE FEDERAL
   MOTOR VEHICLE SAFETY STAN-
   DARDS IN EFFECT ON DATE OF
   MANUFACTURE SHOWN ABOVE.
```

VEH. IDENT NO.		BODY	COL.
1H47F100001			
TRIM	AXLE	TRANS	DSO

NOT FOR TITLE OR REGISTRATION

MADE IN U.S.A

THE VEHICLE IDENTIFICATION NUMBER

The first series of letters and numbers on the rating plate, the VIN number identifies the model year, assembly plant, series, engine and the production sequence.

FIRST DIGIT: Identifies the model year (1971)

SECOND DIGIT: Identifies the assembly plant

ASSEMBLY PLANT	CODE
Atlanta, GA	A
Oakville, CAN	B
Mahwah, NJ	E
Dearborn, MI	F
Chicago, IL	G
Lorain, OH	H
Los Angeles, CA	J
Kansas City, KS	K
Norfolk, VA	N
Twin Cities, MN	P
San Jose, CA	R
Allen Park, MI	S
Metuchen, NJ	T
Louisville, KY	U
Wayne, MI	W
St. Thomas, CAN	X
Wixom, MI	Y

THIRD AND FOURTH DIGITS: Identify the body serial code

TYPE	CODE
Ranchero	46
Ranchero 500	47
Ranchero GT	48
Ranchero Squire	49

FIFTH DIGIT: Identifies the engine

ENGINE	CODE
240 cid, 6 cyl.	V
302 cid, 8 cyl.	F
302 cid, 8 cyl.	6*
351 cid, 8 cyl.	H
351 cid, 8 cyl.	M
351 cid, 8 cyl. GT	Q
429 cid, 8 cyl. CJ	C
429 cid, 8 cyl. CJ Ram-Air	J

LAST SIX DIGITS: Identify the consecutive unit number

THE VEHICLE DATA appears on the two lines following the vehicle identification number.

THE BODY CODE indicates the body style.

BODY STYLE	CODE
Ranchero	66A
Ranchero 500	66B
Ranchero GT	66C
Ranchero Squire	66E

THE EXTERIOR COLOR CODE indicates the paint color used on the vehicle.

COLOR	CODE
Maroon Metallic	B
Dk. Green Metallic	C
Grabber Yellow	D
Prairie Yellow	E
Dk. Vintage Green	G
Seafoam Green	H
Grabber Blue	J
Wimbledon White	M
Diamond Blue	N
Scandia Green Metallic	P
Winter Blue Metallic	Q
Gray Gold Metallic	S
Lt. Pewter Metallic	V
Dk. Blue Metallic	X
Grabber Green Metallic	Z
Bright Red	3
Lt. Gold	8

GLAMOUR PAINTS

COLOR	CODE
Med. Ivy Bronze Metallic	49
Med. Ginger Bronze Metallic	79
Med. Ivy Bronze Metallic	E9
Med. Ginger Bronze Metallic	39
Med. Blue Metallic	D9
Med. Red Metallic	C9

THE INTERIOR TRIM CODE indicates the key to the trim color and material used on the vehicle.

COLOR	CODE
Black knit vinyl	GA,QA,TA,UA
Black vinyl	UA
Med. Blue vinyl	GB
Med. Blue knit vinyl	TB
Med. Vermilion vinyl	GE
Med. Vermilion knit vinyl	TE
Med. Ginger vinyl	GF
Med. Ginger knit vinyl	QF,TF
Med. Green vinyl	GR
Med. Green knit vinyl	QR,TR
White knit vinyl	QW

THE TRANSMISSION CODE indicates the type of transmission installed in the vehicle.

TYPE	CODE
4-Speed manual - wide ratio	5
4-Speed manual - close ratio	6
Automatic (C6)	U

THE REAR AXLE CODE indicates the ratio of the rear axle installed in the vehicle.

RATIO	CONVENTIONAL	LOCK
2.75	2	K
2.79	3	—
2.80	4	M
3.00	6	O
3.25	9	R
3.50	A	S
3.07	B	—
3.91	—	V
4.11	—	Y

THE D.S.O. CODE: Units built on a Domestic Special Order, Foreign Special Order, or other Special orders will have the complete order number in this space. Also to appear in this space is the two-digit code number of the District which ordered the unit. If the unit is a regular production unit, only the Distrit code number will appear.

DISTRICT	CODE
Boston	11
New York	13
Newark	15
Philadelphia	16
Washington	17
Atlanta	21
Charlotte	22
Jacksonville	24
Richmond	25
Louisville	28
Cleveland	32
Detroit	33
Lansing	35
Buffalo	37
Pittsburgh	38
Chicago	41
Milwaukee	43
Twin Cities	44
Indianapolis	46
Cincinnati	47
Denver	51
Dansas City	53
Omaha	54
St. Louis	55
Davenport	56
Dallas	61
Houston	62
Memphis	63
New Orleans	64
Oklahoma City	65
Los Angeles	71
San Jose	72
Salt Lake City	73
Seattle	74
Phoenix	75
Government	83
Home Office Reserve	84
American Red Cross	85
Body Company	87
Transportation Services	89
Export	90's

ENGINE SPECIFICATIONS:

ENGINE CODE	NO. CYL.	CID	HORSE-POWER	COMP. RATIO	CARB
F	8	302	210	9.0:1	2 BC
H	8	351	240	9.0:1	2 BC
M	8	351	285	10.7:1	4 BC
Q	8	351	330	11.7:1	4 BC
K	8	429	370	11.3:1	4 BC
N	8	429	375	11.5:1	4 BC

1972 F-250 RANGER XLT

1972 BRONCO

1972 F-100 PICKUP

1972 F-100 PICKUP

1972 RANCHERO GT

1972 BRONCO

1972 F-100 RANGER XLT

1972 RANCHERO GT

F-SERIES RATING PLATE

The information indicated on the rating plate is the vehicle identification number, the wheelbase, the exterior color, the model type, the body type, the transmission type, the rear axle, the maximum gross vehicle weight (lbs.), the certified net horsepower, the r.p.m. and the D.S.O. numbers.

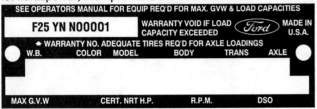

THE VEHICLE IDENTIFICATION NUMBER is a series of letters and numbers on the rating plate. The VIN number identifies the series, engine, assembly plant and the production sequence.

FIRST, SECOND AND THIRD DIGITS: Identify the series

SERIES	CODE
F-100 (4x2)	F10
F-100 (4x4)	F11
F-250 (4x2)	F25
F-250 (4x4)	F26
F-350 (4x2)	F35

FOURTH DIGIT: Identifies the engine

ENGINE	CODE
302 cid, 8 cyl.	G
240 cid, 6 cyl.	A
300 cid, 6 cyl.	B
360 cid, 8 cyl.	Y
390 cid, 8 cyl.	H
240 cid, 6 cyl.	1
300 cid, 6 cyl.	2
302 cid, 8 cyl.	3
360 cid, 8 cyl.	8

FIFTH DIGIT: Identifies the assembly plant

ASSEMBLY PLANT	CODE
Ontario Truck	C
Mahwah, NJ	E
Lorain, OH	H
Kansas City, KS	K
Michigan Truck	L
Norfolk, VA	N
Twin Cities, MN	P
San Jose, CA	R
Allen Park, MI	S
Kentucky Truck	V

LAST SIX DIGITS: Identifies the consecutive unit number

CALENDAR YEAR - 1971	NUMBERS
August 1971 Model	M20,000 thru M29,999
August 1972 Model	M40,000 thru M59,999
September	M60,000 thru M79,999
October	M80,000 thru M99,999
November	N00,000 thru N19,999
December	N20,000 thru N39,999

CALENDAR YEAR - 1972	
January	N40,000 thru N59,999
February	N60,000 thru N79,999
March	N80,000 thru N99,999
April	P00,000 thru P19,999
May	P20,000 thru P39,999
June	P40,000 thru P59,999
July	P60,000 thru P69,000

THE VEHICLE DATA appears on the two lines following the vehicle identification number.

THE W.B. (WHEELBASE) CODE indicates the wheelbase in inches.

THE EXTERIOR COLOR CODE indicates the paint color of the vehicle.

COLOR	CODE
Maroon	2
White	M
Spec. White	C
Calypso Coral	1
Candyapple Red	T
Rangoon Red	J
Med. Blue Metallic	E
Platinum	N
Lt. Blue	B
Med. Blue	7
Med. Blue Metallic	W
Bright Med. Blue	6
Lt. Yellow Green	Q

Med. Ivy Green	K
Prime	9
Lt. Copper Metallic	Z
Med. Bright Aqua	3
Lt. Green	F
Med. Green Metallic	P
Dk. Green	L
Lt. Aqua	R
Med. Turquoise	S
Dk. Green	V
Lt. Yellow	D
Med. Mallard Green Met.	Y
Med. Goldenrod Yellow	K
Chrome Yellow	G
Med. Beige	H
Lt. Ginger Metallic	W
Bright Yellow	5
Med. Metallic Green	O
Med. Metallic Ginger	R

THE MODEL CODE indicates the model type and the gross vehicle weight (lbs.) information.

F-100 (4x2) SERIES

PICKUP	G.V.W.	CHASSIS CAB	G.V.W.
F-100	4,450	F-170	4,450
F-101	4,550	F-171	4,550
F-102	4,800	F-172	4,800
F-103	5,000	F-173	5,000
F-104	5,500	F-174	5,500
F-105*	5,500	F-175*	5,500

F-100 (4x4) SERIES

PICKUP	G.V.W.	CHASSIS CAB	G.V.W.
F-110	5,200	F-180	5,200
F-111	5,600	F-181	5,600
F-112*	5,200	F-182*	5,200
F-113*	5,600	F-183*	5,600

F-250 (4x2) SERIES

PICKUP	G.V.W.	CHASSIS CAB	G.V.W.
F-250	6,200	F-270	6,200
F-251	6,900	F-271	6,900
F-252	7,500	F-272	7,500
F-253	8,100	F-273	8,100
F-254*	7,500	F-274*	7,500
F-255*	8,100	F-275*	8,100

F-250 (4x4) SERIES

PICKUP	G.V.W.	CHASSIS CAB	G.V.W.
F-260	6,500	F-280	6,500
F-261	7,100	F-281	7,100
F-262	7,700	F-282	7,700
F-263*	7,100	F-283*	7,100
F-264*	7,700	F-284*	7,700

F-350 (4x2) SERIES

PICKUP	G.V.W.	CHASSIS CAB	G.V.W.
F-350	6,600	F-370	6,600
F-351	8,000	F-371	8,000
		F-372	8,300
		F-373	9,000
		F-374	10,000
		F-375*	10,000
		F-376*	9,000

* Reference Pennsylvania registration data

THE BODY CODE: The first digit indicates the cab trim, the second the seat type and the third the body code.

CUSTOM	RANGER	RANGER XLT	STD.	COLOR CAB/TRIM
B	K	S	2	Blue
C	L	T	3	Parchment
D	M	U	4	Black
E	N	V	5	Red
F	O	W	6	Green

SEAT CODES

HD BLACK VINYL	DRIVER W/COMP.	SINGLE DRIVER	FULL WIDTH	DESCRIPTION
A	-	-	1	Full width
B	-	-	2	Full width-Custom
-	C	3	-	L-S Unison
-	D	4	-	L-S No. 675
-	E	5	-	Bostrom Westcoaster
-	F	6	-	Bostrom T-Bar
-	G (1)	7	-	National Cush-N-Aire
-	H	8	-	Bostrom Level Air

BODY TYPE

BODY	CODE
Conventional cab	81
Cowl and chassis	84
Cowl and windshield	85

THE TRANSMISSION CODE indicates the type of transmission installed in the vehicle.

TYPE	CODE
New Processs 435 4-speed	A
Warner T-85 overdrive 3-speed	B
Spicer P8516 overdrive	B
Ford-manual 3-speed	C
Warner T-89F 3-speed	D
Clark 387 V 5-speed	D
Warner T-87G 3-speed	E
Fuller 5 H74 5-speed	E
Fuller 5HA74 5-speed	F
Warner T-18 4-speed	F
C-4 automatic	G
Clark 380 overdrive 5-speed	G
Fuller RTO 9513	J
Spicer 6453A 5-speed	K
Allison AT540	L
Clark 285V 5-speed	M
Spicer 6352 5-speed	N
New process 542FL 5-speed	O
Fuller T-905B 5-speed	O
Warner T-19 4-speed	P
Spicer 5652 5-speed	Q
Spicer 8716	R
Spicer 5756-B 5-speed	S
New process 542 FO 5-speed	T
Fuller RTO-9509 B 9-speed	T
Spicer 6852G 5-speed	U
Fuller RT-910	V
Spicer 6352B 5-speed	W
Fuller T-905A 5-speed	X
Transmatic MT-41 6-speed	Y
Transmatic MT-40 6-speed	Z
Spicer 8552A 5-speed	1
Transmatic MT-42 6-speed	1
Clark 282V 5-speed	2
Fuller RT-906	3
Clark 280 Vo 5-speed	4
Fuller RTO-910	5
Fuller RTO-915	6
Clark 385V 5-speed	7
Fuller RT-915	9
New process 542 FD 5-speed	9

THE REAR AXLE CODE: The first two digits represent the rear axle and the third represents the front axle, if applicable.

REAR AXLE CODES

DANA

5250 LB.	CODE
3.54	37
3.73	38
4.10	24
4.56	25

LIMITED SLIP	CODE
DANA 5250 LB.	
3.54	C7
3.73	C8
4.10	B4

3330 LB. FORD	CODE
3.25	17
3.50	08
3.70	09
4.11	05
4.00	02

3300 LB. FORD LIMITED SLIP	CODE
3.70	A2

3600 LB. FORD LIMITED SLIP	CODE
3.50	H2
4.09	H3

7400 LB. DANA 70	CODE
4.10	27
4.56	28
3.73	36
4.88	22

7400 LB. DANA 70 LIMITED SLIP	CODE
4.10	D7

FRONT AXLE CODES

AXLE	CODE
5,000 Ford	A
5,500	B
6,000	C
7,000	D
3,500 Dana 6CF HD	K
3,000 Dana 44 lock	L

THE MAX. G.V.W. LBS. indicates the maximum gross vehicle weight in pounds.

THE CERT. NET H.P. indicates the certified net horsepower at specified rpm.

THE R.P.M. indicates the specified r.p.m. required to develop the certified net horsepower.

THE D.S.O. CODE: If the vehicle is built on a Direct Special order the complete order number will be reflected under the D.S.O. space, including the District Code Number.

DISTRICT CODES

DISTRICT	CODE
Boston	11
Buffalo	12
New York	13
Pittsburgh	14
Newark	15
Philadelphia	16
Washington	17
Atlanta	21
Charlotte	22
Memphis	23
Jacksonville	24
Louisville	28
Chicago	41
Cleveland	42
Milwaukee	43
Indianapolis	46
Cincinnati	47
Detroit	48
Dallas	52
Kansas City	53
Omaha	54
St. Louis	55
Davenport	56
Houston	57
Twin Cities	58
Los Angeles	71
San Jose	72
Salt Lake City	73
Seattle	74
Phoenix	75
Denver	76
Government	83
Home Office Reserve	84
American Red Cross	85
Transportation Services	89
Body Company	87
Export	90's

FORD OF CANADA

Central	B1
Eastern	B2
Atlantic	B3
Midwestern	B4
Western	B6
Pacific	B7
Export	12

ENGINE SPECIFICATIONS:

ENGINE CODE	NO. CYL.	CID	HORSE-POWER	COMP. RATIO	CARB
A	6	240	121	8.5:1	1 BC
B	6	300	165	8.4:1	1 BC
G	6	302	154	8.2:1	2 BC
Y	8	360	196	8.0:1	2 BC
H	8	390	201	8.2:1	2 BC

BRONCO RATING PLATE

The information indicated on the rating plate is the vehicle identification number, the wheelbase, the exterior color, the model type, the body type, the transmission type, the rear axle, the maximum gross vehicle weight (lbs.), the certified net horsepower, the r.p.m. and the D.S.O. numbers.

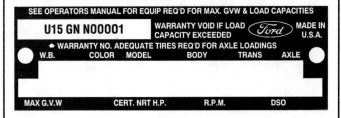

THE VEHICLE IDENTIFICATION NUMBER is a
series of letters and numbers on the rating plate. The VIN number identifies the series, engine, assembly plant and the production sequence.

FIRST, SECOND AND THIRD DIGITS: Identify the series

SERIES	CODE
U-100 pickup	U14
U-100 wagon	U15

FOURTH DIGIT: Identifies the engine

ENGINE	CODE
170 cid, 6 cyl.	F
302 cid, 8 cyl.	G

FIFTH DIGIT: Identifies the assembly plant

ASSEMBLY PLANT	CODE
Lorain, OH	H
Michigan Truck	L
San Jose, CA	R
Allen Park, MI	S

LAST SIX DIGITS: Identify the consecutive unit number

CALENDAR YEAR - 1971	NUMBERS
August - 1971 model	M20,000 thru M29,999
August	M40,000 thru M59,999
September	M60,000 thru M79,999
October	M80,000 thru M99,999
November	N00,000 thru N19,999
December	N20,000 thru N39,999

CALENDAR YEAR - 1972	NUMBERS
January	N40,000 thru N59,999
February	N60,000 thru N79,999
March	N80,000 thru N99,999
April	P00,000 thru P19,999
May	P20,000 thru P39,999
June	P40,000 thru P59,999
July	P60,000 thru P69,000

THE VEHICLE DATA appears on the two lines following the vehicle identification number.

THE W.B. (WHEELBASE) CODE indicates the wheelbase in inches.

THE EXTERIOR COLOR CODE indicates the paint colors used on the vehicle.

COLOR	CODE
Maroon	2
White	M
Spec. White	C
Calypso Coral	1
Candyapple Red	T
Lt. Blue	B
Med. Blue Metallic	W
Bright Med. Blue	6
Lt. Yellow Green	Q
Med. Ivy Green	K
Prime	9
Lt. Copper Metallic	Z
Med. Bright Aqua	3
Lt. Aqua	R
Dk. Green	V
Med. Mallard Green Metallic	Y
Chrome Yellow	G
Med. Metallic Green	O
Tampico Yellow	8
Durango Tan	(Explorer option)

THE MODEL CODE indicates the model type.

SERIES	MODEL
U-100 pickup	U-140
U-100 pickup	U-141
U-100 pickup	U-142
U-100 wagon	U-150
U-100 wagon	U-151
U-100 wagon	U-152

THE BODY CODE: The letter and numerals under body designate the interior trim and body type (the letter identifies the interior trim scheme and the numerals identify the body or cab type.

H.D. VINYL	BRONCO REAR SEAT	STD.	SEATS
B	—	2	Bench seat
C	—	3	Foam cushion
—	4	—	Bronco rear seat

BODY	CODE
Open body (roadster)	U-130
HD package	U-132
Sports Utility	U-140
HD package	U-142
Long roof (wagon)	U-150
HD package	U-152

THE TRANSMISSION CODE indicates the transmission type installed in the vehicle.

TRANSMISSION	CODE
3-speed manual	C

THE REAR AXLE CODE indicates the ratio of the rear axle installed in the vehicle.

AXLE - FORD 2780 LB.	CODE
4.11	03
4.11 lock	A3
4.57	04
3.50	18
3.50 lock	B8

AXLE - FORD 3300 LB.	CODE
4.11 lock	A5
3.50 lock	B9
4.11	05
3.50 lock	08

FRONT AXLE:

AXLE	CODE
Limited slip (Dana 30)	J
Dana 441F	K

THE MAX. G.V.W. LBS. CODE indicates the maximum gross vehicle weight in pounds.

THE CERT. NET H.P. CODE indicates the certified net horsepower at specified r.p.m..

THE R.P.M. CODE indicates the specified r.p.m. required to develop the certified net horsepower.

THE D.S.O. CODE: If the vehicle is built on a Direct Special Order the complete order number will be reflected under the D.S.O. space, including the District Code Number.

DISTRICT	CODE
Boston	11
Buffalo	12
New York	13
Pittsburgh	14
Newark	15
Philadelphia	16
Washington	17
Atlanta	21
Charlotte	22
Memphis	23
Jacksonville	24
Louisville	28
Chicago	41
Cleveland	42
Milwaukee	43
Indianapolis	46
Cincinnati	47
Detroit	48
Dallas	52
Kansas City	53
Omaha	54
St. Louis	55
Davenport	56
Houston	57
Twin Cities	58
Los Angeles	71
San Jose	72
Salt Lake City	73
Seattle	74
Phoenix	75
Denver	76
Government	83
Home Office Reserve	84
American Red Cross	85
Transportation Services	89
Body Company	87
Export	90's

FORD OF CANADA

Central	B1
Eastern	B2
Atlantic	B3
Midwestern	B4
Western	B6
Pacific	B7
Export	12

ENGINE SPECIFICATIONS:

ENGINE CODE	NO. CYL.	CID	HORSE-POWER	COMP. RATIO	CARB
F	6	170	97	8.3:1	1 BC
G	8	302	159	8.2:1	2 BC

RANCHERO
RATING PLATE

The information indicated on the rating plate is the vehicle identification number, the body type, the exterior color, the interior trim, the rear axle, the transmission type and the D.S.O. numbers.

MANUFACTURED BY   100001
FORD MOTOR COMPANY
08/70 THIS VEHICLE CONFORMS
TO ALL APPLICABLE FEDERAL
MOTOR VEHICLE SAFETY STAN-
DARDS IN EFFECT ON DATE OF
MANUFACTURE SHOWN ABOVE.

VEH. IDENT NO.	BODY	COL.
2H47F100001		

TRIM	AXLE	TRANS	DSO

NOT FOR TITLE OR REGISTRATION

MADE IN U.S.A

THE VEHICLE IDENTIFICATION NUMBER is a series of letters and numbers on the rating plate. The VIN number identifies the model year, assembly plant, series, engine and the production sequence.

FIRST DIGIT: Identifies the model year (1972)

SECOND DIGIT: Identifies the assembly plant

ASSEMBLY PLANT	CODE
Atlanta, GA	A
Oakville, CAN	B
Mahwah, NJ	E
Dearborn, MI	F
Chicago, IL	G
Lorain, OH	H
Los Angeles, CA	J
Kansas City, KS	K
Norfolk, VA	N
Twin Cities, MN	P
San Jose, CA	R
Allen Park, MI (Pilot)	S
Metuchen, NJ	T
Louisville, KY	U
Wayne, MI	W
St. Thomas, CAN	X
Wixom, MI	Y
St. Louis, MO	Z

THIRD AND FOURTH DIGITS: Identify the body serial code

BODY SERIAL CODE	CODE
Ranchero 500	47
Ranchero GT	48
Ranchero Squire	49

FIFTH DIGIT: Identifies the engine

ENGINE	CODE
250 cid, 6 cyl.	L
250 cid, 6 cyl.	3*
302 cid, 8 cyl.	F
302 cid, 8 cyl.	6*
351 cid, 8 cyl.	H
351 cid, 8 cyl.	Q
351 cid, 8 cyl.	R
400 cid, 8 cyl.	S
429 cid, 8 cyl.	N

* Low compression export

LAST SIX DIGITS: Identify the consecutive unit number

THE VEHICLE DATA appears on the line following the vehicle identification number.

THE BODY CODE indicates the body style.

BODY STYLE	CODE
Ranchero 500	97D
Ranchero GT	97R
Ranchero Squire	97K

* Ranchero Special on Ranchero 500 only

THE EXTERIOR COLOR CODE indicates the paint colors used on the vehicle.

COLOR	CODE
Lt. Gray Metallic	1A
Black	1C
Silver Metallic	1D
Calypso Blue	2A
Bright Red	2B
Red	2E
Med. Red Metallic	2G
Maroon	2J
Lt. Blue	3B
Med. Blue Metallic	3C
Med. Blue Metallic	3D
Grabber Blue	3F
Dk. Blue Metallic	3H
Bright Blue Metallic	3J
Pastel Lime	4A
Bright Green Gold Metallic	4B
Ivy Bronze Metallic	4C
Med. Ivy Bronze Metallic	4D
Med. Lime Metallic	4F
Med. Ivy Bronze Metallic	4G
Med. Green Metallic	4P
Dk. Green Metallic	4Q
Lt. Pewter Metallic	5A
Ginger Bronze Metallic	5C
Ginger Bronze Metallic	5D
Dk. Brown Metallic	5F
Lt. Copper Metallic	5G
Ginger Metallic	5H
Med. Ginger Bronze Metallic	5J
Tan	5L
Lt. Goldenrod	6B
Med. Goldenrod	6C
Yellow	6D
Med. Bright Yellow	6E
Bright Yellow Gold Metallic	6F
Lt. Copper Metallic	6G
Gray Gold Metallic	6J
White	9A
Gold Glow	*
Ivy Glow	*

* On Ranchero Special

THE INTERIOR TRIM CODE indicates the key to the trim color and material used on each model.

COLOR	VINYL	CLOTH	LEATHER	CODE
Black	•			AA,CA,FA
Black	•			GA,RA,UA
Black	•	•		BA,DA,EA
Black	•	•		VA,HA,JA
Black	•	•		MA,KA,NA
Black	•	•		QA
Med. Blue	•			AB,CB,FB
Med. Blue	•			GB,RB,UB
Med. Blue	•	•		BB,KB,MB
Med. Blue	•	•		NB,QB
Med. Green	•			AR,CF,CR
Med. Green	•			FR,GR,RR
Med. Green	•			UR
Med. Green	•	•		BR,KR,MR
Med. Green	•	•		NR,QR
Lt. Gray Gold	•	•		BY,NY,QY
Tobacco	•	•		BZ,NZ,QZ
Ginger	•			CF,FF,GF
Ginger	•	•		DF,EF,VF
Ginger	•	•		HF,MF,JF
Ginger	•	•		KF,MF
White/Black	•			CW,FW,RW
White/Black	•			UW
White/Black	•	•		KW,MW
White/Blue	•			CL,FL,RL
White/Blue	•			UL
White/Blue	•	•		KL,ML
White/Tobacco	•			C9,F9,R9
White/Tobacco	•			U9
White/Tobacco	•	•		K9,M9
White/Green	•			C5,F5,R5
White/Green	•			U5
White/Green	•	•		K5,M5

THE REAR AXLE CODE indicates the ratio of the rear axle installed in the vehicle.

REAR AXLE	CODES CONVENTIONAL	LOCK
2.75	2	K
2.79	3	—
2.80	4	M
3.00	6	O
3.18	7	—
3.25	9	R
3.50	A	S
3.07	B	—
3.55	G	—
3.78	H	—
3.91	—	V

THE TRANSMISSION CODE indicates the type of transmission installed in the vehicle.

TYPE	CODE
3-Speed manual	1
4-Speed manual	5
4-Speed manual	E
Automatic (C4)	W
Automatic (C6)	U
Automatic (FMX)	X
Automatic (C6 Special)	Z

THE D.S.O. CODE indicates the DSO code, consisting of two numbers, designating the district in which the car was ordered and may appear in conjunction with a Domestic Special Order or Foreign Special Order number when applicable. Ford of Canada DSO codes consist of a letter and a number.

DISTRICT	CODE
Boston	11
Buffalo	12
New York	13
Pittsburgh	14
Newark	15
Philadelphia	16
Washington	17
Atlanta	21
Charlotte	22
Memphis	23
Jacksonville	24
Richmond	25
New Orleans	26
Louisville	28
Chicago	41
Cleveland	42
Milwaukee	43
Lansing	45
Indianapolis	46
Cincinnati	47
Detroit	48
Dallas	52
Kansas City	53
Omaha	54
St. Louis	55
Davenport	56
Houston	57
Twin City	58
Los Angeles	71
San Jose	72
Salt Lake City	73
Seattle	74
Phoenix	75
Denver	76
Government	83
Home Office Reserve	84
American Red Cross	85
Body Company	87
Transportation Services	89
Export	90's

ENGINE SPECIFICATIONS:

ENGINE CODE	NO. CYL.	CID	HORSE-POWER	COMP. RATIO	CARB
L	6	250	99	8.0:1	1 BC
F	8	302	141	8.5:1	2 BC
H	8	351	164	8.6:1	2 BC
Q	8	351	248	8.6:1	4 BC
S	8	400	168	8.4:1	2 BC
N	8	429	205	8.5:1	4 BC
P	8	429	212	8.5:1	4 BC